DATE DUE

THEOLOGY AND SCIENCE IN THE
THOUGHT OF FRANCIS BACON

To Stephen A. McKnight,
with gratitude

Theology and Science in the Thought of Francis Bacon

STEVEN MATTHEWS
The University of Minnesota, USA

ASHGATE

Published by
Ashgate Publishing Limited
Gower House
Croft Road
Aldershot
Hampshire GU11 3HR
England

Ashgate Publishing Company
Suite 420
101 Cherry Street
Burlington, VT 05401-4405
USA

Ashgate website: http://www.ashgate.com

British Library Cataloguing in Publication Data
Matthews, Steven, 1969–
   Theology and science in the thought of Francis Bacon
   1. Bacon, Francis, 1561–1626 2. Religion and science
   3. Theology
   I. Title
   192

Library of Congress Cataloging-in-Publication Data
Matthews, Steven, 1969–
   Theology and science in the thought of Francis Bacon / by Steven Matthews.
       p. cm.
   Includes bibliographical references.
   ISBN 978-0-7546-6252-5 (alk. paper)
1. Bacon, Francis, 1561–1626. 2. Religion and science. 3. Theology. I. Title.

   B1198.M38 2008
   192–dc22

                                                                                2007032297

ISBN 978-0-7546-6252-5

Printed and bound in Great Britain by TJ International Ltd, Padstow, Cornwall.

Contents

Preface

What is the reason for yet another book on Francis Bacon? Many scholars have specialized in studying the thought and influence of this one individual, and it may seem as if we now. over a distance of four centuries, know him about as well as we possibly can. Yet when we compare the hundreds of books and thousands of articles written on Bacon's life and thought, very different images of Bacon emerge. To some degree, this is merely a reflection of the varied interests of scholars and authors who study him as a statesman, a philosopher, an author, or the shadowy figure behind a conjectured theatrical conspiracy. Another cause of the many different Bacons in the secondary literature is a lack of scholarly consensus on how Bacon related to the dominant cultural force of his age, reformation era Christianity. Bacon has been portrayed as an atheist, a Puritan, a generic "sincere Christian," and one who, whatever his religious beliefs, was unconcerned with matters of faith in his philosophy.

The question of Bacon's faith is especially significant for understanding his philosophy and his cultural role as an acknowledged "founding figure" of the modern scientific method and worldview. Bacon's writings pertaining to natural philosophy and the reform of human learning, the program which he entitled the *Instauratio Magna*, are saturated with scriptural quotations and theological arguments used to support his points. Are these statements sincere? Are they a twisting of Christianity for his secular ends? Could they merely have been inserted as window dressing to please a Christian readership? These ideas and many more have emerged to explain the religious language of Bacon's writings, but almost no attention has been given to interpreting them within the historical context of the theological trends of the Reformation, where any answers must be found. The questions are certainly important for understanding Bacon, for his philosophy, undergirded as it is with religious arguments and Scripture, takes on radically different implications if it is assumed to be written by a Calvinist, an atheist, or someone who was simply trying to keep "faith" and "science" separate. It has been my concern in this book to place Bacon back in his proper day and age, and let his own writings inform us about where he fitted in the theological landscape of Tudor and Stuart England.

Most of the conflicting images of Bacon's religion have emerged because they reflect the concerns and interests of later generations, and not those of Bacon's own era. John Henry has observed that the portrayal of Bacon as an atheist or a deist at odds with Christianity cannot be found prior to the Enlightenment.[1] Indeed, it cannot.

It may be that the Enlightenment admirers of Bacon recast him in their own image. Perhaps this misreading of Bacon can be placed more properly on the shoulders of Joseph de Maistre, one of the Enlightenment's fiercest early critics who, in the wake of the problems of the French Revolution, blamed Bacon for just

---

[1]   John Henry, *Knowledge is Power: How Magic, the Government and an Apocalyptic Vision Inspired Francis Bacon to Create Modern Science* (London, 2002), p. 83.

about everything, but especially for the atheism which led to the fall of crown and altar. In any case, such a reading of Bacon renders every single religious statement made by Bacon hypocritical (including his prayers and his personal *Confession of Faith*). This is doubtful enough, but there is an additional problem which arises when we shift the focus from Bacon to his rather small circle of close friends. This was a very pious group, if not very typical of the dominant English Calvinism of the period. It included Bishop Lancelot Andrewes, Father Tobie Matthew, and Bacon's chaplain, William Rawley, among others. Were these men in on the conspiracy to hide Bacon's anti-Christian agenda, or were they unaware of what would be so clear to later Bacon scholars? There are serious historical problems with this position, and it is reasonable to conclude that it stems more from issues which Joseph de Maistre had with the Enlightenment than from Bacon himself.

Nevertheless, the reading of Bacon as an atheist persists today in scholarship which casts Bacon as a Machiavellian manipulator who sought to undermine traditional Christianity. There can be no question that Bacon was an admirer of Machiavelli, or that he was as shrewd a political operator as could be found at court. But all of this does not preclude him from being a true believer in Christianity. There were plenty of manipulative Christians in the Tudor and Stuart eras, and they were no less sincere for playing the political game. Their beliefs often served as the ultimate motive for their maneuvers. This was the case with Francis Bacon.

A more common reading of Bacon today presents him as a sincere Christian of one variety or another, but one who insisted that somehow "science" and "faith" should be kept separate. Beyond the inherent anachronism of discussing "science" in the age of natural philosophy, this position raises the question of why Bacon spent so much ink and effort in discussing theology in his philosophical writings if he was trying to effect such a separation. This position can only be maintained through a selective reading of Bacon which must overlook, among other things, a large section in *De Augmentis Scientiarum* which sets forth a plan for the reform of theology along with all of the other sciences (using the original meaning of the term – "human learning" – which was not limited to natural philosophy). There are passages in Bacon's writing which appear to lead toward a separation of science and theology, particularly if they are read in English translation rather than in their original Latin. Again, the idea of this separation is really a later development. It reflects the interests of certain (though not all) of Bacon's disciples in the Royal Society, who were concerned with avoiding the religious conflicts which had contributed to the Civil War.

Over the last ten to twenty years there has also been a trend toward understanding Bacon as a creature of his context, and this implies taking his theological statements seriously and interpreting them in light of the religious fabric of Tudor and Stuart England. A host of nuanced studies have contributed to a more historically accurate image of Bacon and his thought, and many of these may be found in my notes and my bibliography. In spite of this increase in scholarly activity Stephen McKnight has very accurately observed that "there is still no book-length analysis of Bacon's use of religious images and themes in his major works, and there is no systematic development of Bacon's religious outlook."[2]

---

[2]   Stephen A. McKnight, *The Religious Foundations of Francis Bacon's Thought* (Columbia, 2006), p. 2.

McKnight's own book addressed the first of these lacunae, and my book is aimed at the second, the "systematic development of Bacon's religious outlook."

Bacon's context is the key to my approach, and the first chapter establishes the features of the English Reformation which shaped Bacon's thought, and places Bacon within this context. The second chapter traces Bacon's trajectory away from the Puritanism of his mother and toward an eclectic anti-Calvinism which can be seen in his most developed statements on religion. Chapters 3, 4, and 5 examine the features of Bacon's theology as they inform his writings pertaining to the reform of learning and the Great Instauration. These chapters include a special consideration of how "theology" and "natural philosophy" form a single system of thought for Bacon. Chapter 6 examines Bacon's literary circle as a further context for understanding Bacon, and considers how Bacon's image was transformed to fit later interests and agendas.

Steven Matthews
The University of Minnesota, Duluth

Acknowledgments

No list of acknowledgments for a project such as this can ever be complete. Certain people and organizations, however, require special mention. Among the first must be the history department of the University of Florida, and especially Stephen McKnight, Fred Gregory, Robert Hatch, and C. John Sommerville. Their cordial and candid discussions with me are responsible for most of the positive directions my research has taken.

A special thank you goes to those who have commented on the final drafts of this text: Alexis Pogorelskin, Kyle Dibb, John Cirilli, and especially Bryn Johnson, who, in the words of Bacon "was my inquisitor."

The financial support of the Earhart Foundation, as well as the Intercollegiate Studies Institute, must also be acknowledged. Their generous support during the formative years of this project permitted me to turn my time and my other resources toward research. A Francis Bacon Foundation fellowship at the Huntington Library ensured that the project was placed on a solid footing, and special thanks must go to the staff of the Huntington Library who assisted me during my several visits there. They are the most gracious and helpful people one may have the opportunity to work with. Thanks also to Linda Krug, and the College of Liberal Arts at the University of Minnesota, Duluth, who ensured that I had the support to finish the project.

Finally, I must acknowledge the love and support of my family, and especially of my wife, Nancy, and my sons, Andrew and Ian. Thank you for your patience, your understanding, and your encouragement.

List of Abbreviations

ANF   Roberts, Alexander and Donaldson, James (eds) *The Ante-Nicene Fathers* (10 vols, Grand Rapids: Hendrickson, 1994). Reprint of the Edinburgh edition.

DNB   Smith, George, Stephen, Sir Leslie, and Lee, Sidney (eds), *Dictionary of National Biography* (21 vols, London: Oxford University Press, 1882–1900).

ICR   Calvin, John, *Institutio Christianae Religionis*. Latin quotations are taken from the London edition of 1576, which was available to Bacon. English translations are primarily from *Institutes of the Christian Religion*, trans. Ford Lewis Battles (Philadelphia: Westminster Press, 1960). Other translations (as noted) are either my own or from the Beveridge translation (Edinburgh, 1845).

NPNF   Schaff, Philip, and Wace, Henry (eds), *The Nicene and Post-Nicene Fathers* (2 series, 14 vols each, Grand Rapids: Hendrickson, 1994). Reprint of the Edinburgh edition.

WFB   Spedding, James, Ellis, Robert Leslie, and Heath, Douglas Denon (eds.), *The Works of Francis Bacon* (14 vols, London, 1858–74).

Chapter 1

Breaking with a Puritan Past

A Mother's Concern

Anne Bacon was a woman of the godly sort. She had been raised to it. Her father, Sir Anthony Cooke, had educated her in Latin as well as in the Greek of the New Testament and the Church Fathers.[1] His own belief was profound. He had been tutor and advisor to the young Edward VI, and when the reign of Mary Tudor began in 1554, he became one of the "Marian exiles" who fled to Geneva. For families such as that of Anthony Cooke, Geneva was more than a place of refuge. In the city of John Calvin all Christians of the Reformed branch of Protestantism could experience the ultimate model of a faithful society. The people of Geneva, by their own account, had not stopped short of a full Reformation as had the Lutherans, who retained many popish ceremonies and doctrines. Neither had the Reformation derailed here as it had among the Anabaptists, who, according to the Reformed, had invented many of their own doctrines, and whose communities could spiral into chaos and anarchy. In Geneva the Law of God as found in the Bible, and only that Law, reigned supreme, resulting in order and genuine piety. For those English men and women who had gone into exile, Geneva was the measure by which their own society was to be judged, and it always fell short. Anne shared the views of her father and the other exiles, but she had stayed behind as she had recently married Sir Nicholas Bacon.

Anne Bacon had two sons whom she had tried to raise the right way. She had seen to much of their education herself, and had hired the very best tutors. The education of her children was of the utmost importance to Anne, for her boys were growing up with unwholesome influences all around. Although England had officially become Protestant once again with the death of Queen Mary and the accession of Elizabeth, the Elizabethan Religious Settlement had been a profound disappointment to Anne and many others like her, for the queen had sought peace through compromise and the toleration of a wide variety of opinions rather than completing the Reformation. The roots of the Puritan movement are to be found among the families of the Geneva exiles, and this can certainly be seen in the convictions of Anne Bacon. She rebelled against the guidelines laid down by Elizabeth and her bishops, and she prayed for a purer Church. Anne had done her best with Anthony and his younger brother, Francis, but, once they were grown, she had become convinced that something had gone wrong somewhere with Francis.

In 1592 Anthony Bacon, then thirty-four years old, received a letter from his mother, filled with a mother's concern over her son's spiritual well-being. The letter

---

[1]   On the life of Anne Bacon see Lisa Jardine and Alan Stewart, *Hostage to Fortune: The Troubled Life of Francis Bacon* (New York, 1998).

was prompted, at least in part, by the apparent wandering of thirty-one-year-old Francis. The following is an excerpt:

> This one chiefest counsel your Christian and natural mother doth give you even before the Lord, that above all worldly respects your carry yourself ever at your first coming as one that doth unfeignedly profess the true religion of Christ, and hath the love of the truth now by long continuance fast settled in your heart, and that with judgment, wisdom, and discretion, and are not afraid or ashamed to testify the same by hearing and delighting in those religious exercises of the sincerer sort, be they French or English. *In hoc noli adhibere fratrum tuum ad consilium aut exemplum. Sed plus dehinc.* [In this do not be willing to consult your brother in counsel or example, but more on this to follow.] If you will be wavering (which God forbid, God forbid), you shall have examples and ill encouragers too many in these days, and that αρch Βισσ, since he was Βουλευτὴς, ἐστὶ ἀπολεία τῆς ἐκκλησίας μεθ' ἡμῶν· φιλεῖ γὰρ τὴν ἑαυτοῦ δόξαν πλέον τῆς δόξης τοῦ Χριστοῦ [and that arch Bish(op), since he was a councilor, is the destruction of the Church among us, since he is more fond of his own glory than the glory of Christ].[2]

Anne is alluding to a number of points which would have been common knowledge to Anthony and herself. Those parts of her letter which might cause real scandal, if they were to be casually seen and become a matter for gossip, have been given a greater measure of secrecy by placing them in Latin and Greek. Nevertheless, the message is clear from a historical context. The archbishop mentioned, for example, is Whitgift, who had by this time a well-earned reputation as the scourge of the Puritan movement. Anne's words, "but more to follow," point toward a postscript in the letter in which she tells Anthony that she assumes his entire household is gathering for prayer twice a day, "having been where reformation is" (since Anthony had been to Geneva and the Calvinist lands on the continent). She follows this with the remark, "your brother is too negligent herein." Apparently, the example which Anthony was not to follow was that of his brother who was not observing proper Calvinist patterns of personal devotion. At the very least, Francis was too tolerant of those in his household who did not follow these patterns. But there was more to Lady Anne's concern than merely a haphazard prayer life. Anthony is warned against consulting his brother's "counsel." Francis was not only lax, he had opinions which, to Anne's thinking, ran counter to "the true religion of Christ."

Francis Bacon's own writings from this time tell much of the rest of the tale. There is a recognizable trajectory in Bacon's adult life away from his Puritan upbringing, and ultimately away from the dominant Calvinism of his society as well. But this was also a trajectory *toward* some of the other options available to him in late Tudor England. Bacon's abandonment of his mother's Christianity does not mean that he himself abandoned Christianity. As with so many in the Reformation era, he wanted to get Christianity right. Over the last decade of the sixteenth century Bacon wrestled with the various theological issues and ideas that were current in his society. What resulted was a coherent belief system, unique to him in many points, which suffused his writings pertaining to the reform of learning and the advancement of sciences, or his "Great Instauration." Bacon's entire understanding of what we call "science," and what he called "natural philosophy," was fashioned around the basic tenets of his belief

---

[2]   WFB, vol. VIII, p. 112.

system. Understanding Bacon's early rejection of Calvinism is key to understanding the Instauration writings themselves, particularly those mysterious passages where Bacon suddenly breaks into biblical quotation and theological discourse. So, before we can proceed to Bacon's writings it is important that we take a good look at the intellectual environment of late Tudor England, which could give rise to a godly mother's concern.

Turmoil and Diversity in the English Reformation

The Reformation was an era of intellectual turmoil and a remarkable diversity of thought which too often has been told as a tale of Protestants and Catholics. However, there was always more on the table than such a simple dichotomy would suggest. "Protestantism" never existed as a unified set of beliefs in the sixteenth century. The Lutherans, the Reformed, and the Anabaptists were all technically "Protestant," yet their differences with each other were as great as the differences of each with the Church of Rome. From the Catholic side, the apparent doctrinal unity of the Council of Trent always cloaked tremendous latitude in interpretation and practice. The Reformation on the continent was far from a tidy affair, characterized by debate even within unified movements such as Lutheranism or the Reformed, but the situation in Bacon's England was even more complex.

Much of the turmoil and diversity of England at the time can be understood as what must happen when a Catholic king and "Defender of the Faith" finds it suddenly necessary to break with the Church of Rome for reasons other than religion. In this case the need was for a divorce from Catherine of Aragon, and Henry, for his part, was not particularly interested in doctrinal changes beyond those that would permit the divorce. Compared with the Reformation on the Continent, the English Reformation was effected out in reverse: first came the break with Rome, and the theology necessary to justify that break and establish a new ecclesial order followed.

It is important to note that what set the English Reformation apart was not that it was an act of state rather than Church.[3] At some point, the Reformation was always an act of state, as the decree of the ruler was necessary to safeguard the existence of non-Roman Christianity everywhere. The principle of the Peace of Augsburg that the ruler should determine the religion of a province (*cuius regio ejus religio*) was, in many respects, not an innovative idea, but an acknowledgment of the way in which the Reformation had developed ever since Elector Frederick of Saxony gave Luther his protection. In Scandinavia, as in England, the Reformation occurred through specific decrees of kings. Throughout Scandinavia the Reformation occurred from above, and for reasons which were far from purely religious.[4] But in these countries, unlike England, the doctrinal choice was clear: Roman doctrine was being rejected

---

[3]   There is an unfortunate debate within the historiography of the English Reformation over whether the Reformation was primarily an act of state or a religious development. For a summary of the basic ideas involved see J.F. Davis, "Lollardy and the Reformation in England" in Peter Marshall (ed.), *The Impact of the English Reformation: 1500–1640* (London, 1997). pp. 37–52. All reformations were both acts of state and religious, if they succeeded at all.

[4]   For a brief account which balances political motives with the religious interests of the Lutheran movement see Harold J. Grimm, *The Reformation Era 1500–1650* (New York, 1954), pp. 235–41.

in favor of the doctrine of Lutheranism, as clearly stated in the Augsburg Confession and the mass of writings streaming northward from Wittenberg. Subscription of the Augsburg Confession meant adoption of the Lutheran package *in toto*.

In the Palatinate and in those cantons of Switzerland which adopted the Reformed faith the theological formulations were also clear, although initially a single normative statement such as the Augsburg Confession was often lacking. The distinction between Lutheran and Reformed was established along specific doctrinal lines by the reformers themselves, and although the idea of confessional subscription did not function so rigidly in the Reformed lands, conformity to Reformed doctrine was expected. After some early disputes Geneva adhered to the doctrine of Calvin, and those who did not adhere were welcome to leave (with some exceptions, of which Servetus is the most notable). Henry VIII had no such doctrinal agenda, however, when he broke with Rome.

In some measure, the king himself had blocked the possibility of confessional unity in England. Until 1536 Henry's actions were designed to transfer decision-making power from the Roman Catholic authorities to himself. The Ten Articles which were forwarded in 1536 as a doctrinal statement were ambiguous by design. They left room for both Catholic and Lutheran interpretations, though between a Lutheran and a Catholic, the Catholic probably would have been more comfortable with them, given their interpretation of sacraments and tradition. Rather than a positive doctrinal statement, A.G. Dickens observed that the Ten Articles might "be used to exemplify our English talent for concocting ambiguous and flexible documents."[5] In the *Bishops' Book* of the following year, the doctrinal position is still more Catholic, but subscription was never enforced, and Henry "used it instead to test the theological appetite of the nation."[6] Catholic doctrine was not particularly discouraged, beyond the question of allegiance to Rome. At the same time, the break with Rome encouraged the development of nascent Protestant movements in England, and these movements were fueled by the appearance of Protestantism which came with the dissolution of monasteries and the seizure of Church property. The replacement of Catholic bishops with Lutheran superintendents and the enforced subscription of the Augsburg Confession, which made it possible for Scandinavian kings to obtain rapid uniformity, had no parallel in England. The wholesale adoption of Wittenberg's pattern of doctrine and liturgy which occurred under the kings of Sweden and Denmark was not possible for Henry, not the least because he had distinguished himself early on as an enemy of Luther. Alec Ryrie has aptly summarized Henry's problem with a Lutheran solution which would have brought swift uniformity:

> As [Basil] Hall has argued, the king's suspicion of Lutheranism in general, his loathing of Luther in particular, and his heartfelt attachment to his own authority guaranteed that the English Church would remain beyond Wittenberg's sphere of influence. Henry's reformation was, as Richard Rex has recently emphasized, 'its own thing, folly to Catholics and a stumbling block to protestants.'[7]

---

[5] A.G. Dickens, *The English Reformation* (second edition, University Park, 1991), p. 200.

[6] Ibid.

[7] Alec Ryrie, "The Strange Death of Lutheran England," *The Journal of Ecclesiastical History*, 53/1 (January 2002), pp. 66–7. Ryrie also points out that any tendency toward Lutheranism as a settlement among the English Protestants themselves was thwarted both by

Throughout Henry's reign, the Church of England remained a Church without a doctrinal identity. The long-term effect was to allow a tremendous doctrinal diversity to develop.

The Thirty-Nine Articles, when they came about, did provide genuine stability and unity for the Church of England, but this stability and unity should not be confused with any great degree of doctrinal uniformity. Rather, they should be recognized as allowing and establishing tremendous doctrinal latitude within the official Church during this era. The accession of Elizabeth and the actions of queen and parliament up through the Act of Uniformity of 1559 established the Church of England as genuinely Protestant, and the official adoption of the Thirty-Nine Articles at this time (in 1562 by Convocation and by parliament in 1571) was the part of that stabilizing chain of events which addressed doctrine directly. Yet it has often been noted that the most remarkable feature of the Thirty-Nine Articles is their ambiguity, which stems partly from the mixture of Lutheran and Calvinist sources in their composition.[8] Although the wording of articles on predestination and the Lord's Supper is typical of Calvinist formulations, there is no requirement that these articles be interpreted according to Calvinist doctrine. Attempts to refine the meaning of the Thirty-Nine Articles by incorporating the Lambeth Articles and rendering the interpretation to be unequivocally Calvinist were rejected both by Queen Elizabeth and later, King James. With careful reading, the articles could be, and were, interpreted from almost every Protestant angle. They are ambiguous enough to have been embraced by both a committed Calvinist, Archbishop Whitgift, and a committed anti-Calvinist, Archbishop Laud.[9]

Another reality of the Elizabethan Settlement was that it would not be, nor could be, thoroughly enforced. Neither Elizabeth nor Lord Chancellor Burghley was interested in tactics that would be seen by the queen's subjects as religious persecution. The only group which could claim martyrdom under Elizabeth by the end of her reign would be the Roman Catholics, and action was only taken against them when it was clear that some of their number were actively working to subvert the realm.[10] The true goal of the Religious Settlement, including the Thirty-Nine Articles, was not doctrinal uniformity but national unity and an orderly and peaceful realm. In application, attempts at forcing uniformity often backfired at the local level, leading bishops to turn a blind eye to religious diversity rather than provoke a reaction.[11] King James continued Elizabeth's policy of promoting a broad and

---

reaction against the king driving Protestant divines toward a more radical position, and by the complicating factor of native Lollardy (pp. 85–92). Without Lutheranism being imposed from above, there was already too much diversity among the anti-Roman Catholics themselves for "Lutheran moderation" to be a real option.

[8] Philip Schaff describes Cranmer's use of Lutheran and moderately Calvinist sources in *History of the Christian Church* (8 vols, New York, 1910), vol. 8, p. 817.

[9] On Laud see John F.H. New, *Anglican and Puritan: The Basis of Their Opposition, 1558–1640* (Stanford, 1964), p. 75. The term "anti-Calvinist" is from Nicholas Tyacke, *Anti-Calvinists: The Rise of English Arminianism, 1590–1640* (Oxford, 1987).

[10] Dickens, *English Reformation*, p. 382.

[11] Claire Cross, *Church and People, 1450–1660: The Triumph of the Laity in the English Church* (Atlantic Highlands, 1976), pp. 124–53. Her discussion demonstrates that doctrinal uniformity was, as a rule, sacrificed for unity.

tolerant Protestantism.[12] For King James, just as for Queen Elizabeth, theological squabbles were the lesser threat, and alienating large numbers of his subjects the greater. The consequences of forcing controversy underground would be manifest only in the reign of Charles I, with the Archiepiscopacy of Laud.

The Thirty-Nine Articles lacked the normative control of the Augsburg Confession among the continental Lutherans, whose ministers were often removed if they disagreed with any aspect of the document, and hence the Articles failed to achieve that level of confessional unity. Similarly, they lacked the common popular assent and enforceable authority of the *Institutes*, *The Geneva Confession*, and the *Heidelberg Catechism* in the Reformed lands. England's religious diversity was beyond the point where it could still be reined in with a demand for confessional subscription. However, the ambiguity of the Thirty-Nine Articles served the agenda of national unity well, while recognizing that the Church of England was too diverse for rigid doctrinal unity.

The actual diversity of religious thought in Tudor and Stuart England is poorly represented by the traditional continuum which places Catholics at one end, Puritans at the other, and the official state Church between the two as a "*via media*." This three-part continuum remains popular today, and it is not without the legitimation of historical precedent: it was used even in the Tudor era as a convenient way of simplifying the religious disputes at the time. The problem with the continuum is that, although it has always been convenient, it has never been accurate.

In 1603, at the dawn of the Stuart era, a tract of polemical verse was published entitled, "The Interpreter, wherein three principal Terms of State, much mistaken by the vulgar, are clearly unfolded." Significantly, the tract demonstrates both the prominence of the three-part continuum at the time and the early recognition that there were problems with it. The poem begins as follows:

> Time was, a Puritan was counted such
> As held some ceremonies were too much
> Retained and urged; and would no Bishops grant,
> Others to Rule, who government did want.
> Time was, a Protestant was only taken
> For such as had the Church of Rome forsaken;
> Or her known falsehoods in the highest point:
> But would not, for each toy, true peace disjoint.
> Time was, a Papist was a man who thought
> Rome could not err, but all her Canons ought
> To be canonical: and, blindly led,
> He from the Truth, for fear of Error, fled.
> But now these words, with divers others more,
> Have other senses than they had before:
> Which plainly I do labour to relate,
> As they are now accepted in our state.[13]

In the rest of the tract the tidy definitions which operated for Puritans, Protestants, and Papists once upon a time are presented as having become hopelessly complicated by

---

[12]   Ibid., p. 153.

[13]   Charles H. Firth (ed.), *Stuart Tracts: 1603–1693* (Westminster, 1903), p. 233.

political interests. The author shows a marked preference for the Puritans, who act as a magnet for those focused on the noble cause of the Reformation. The worst that can be said of them, according to the author, is that their dedication to higher principles means that they may not recognize what is in the best interest of the state. The other two groups, the Protestants (the typical *via media*) and the Papists, are presented as having betrayed their doctrinal convictions for political loyalties, the former to the English king and the latter to his overthrow. It is certain that the "Protestant" or the "Papist" would not share the author's appraisal of what went wrong with the categories. Nevertheless, the tract is a witness to the antiquity of the threefold division and the early recognition of the instability of the categories.

Current scholarship on the English Reformation reveals even less validity in the threefold continuum, with no evidence for the tract author's assumption of an original period of tidy doctrinal dividing lines. "Catholic" is, of course, the easiest category to define in Tudor and Stuart England, if we allow that it requires allegiance to the pope, adherence to a body of traditional doctrines, and eventual adherence to the doctrine of the Council of Trent. This should not be understood to mean that it was by any means easy to tell who was and who was not a Catholic. On the one hand, when open adherence to Catholicism was equated with treason, English Catholicism went underground. On the other hand, accusations of Catholicism were freely made (and sometimes with a degree of accuracy) to tarnish the image of those regarded as not adequately Protestant. Nonetheless, the term "Catholic" signifies a coherent doctrinal identity which is lacking among the other two "typical" groups of the English Reformation.

The "Puritan" mentioned in the tract above was also a recognized category at the time, and the term was not merely, as some have claimed, a pejorative label applied to those who were considered too radical in their Protestantism.[14] While some objected to the label, others such as the tract writer regarded it as a mark of having one's priorities straight. Nevertheless, a single, coherent definition of what Puritans believed is difficult to establish.

The approach of Patrick Collinson has come to form a basis for consensus in the historical study of Puritanism. In his landmark treatment of the subject, *The Elizabethan Puritan Movement*, "puritanism" is acknowledged to be just as "loosely defined" and "widely dispersed" as the various uses of the appellation at the time would indicate.[15] Yet, in the course of the Elizabethan era Collinson discerns the rise of a "puritan movement" with a discreet and recognizable agenda that would manifest itself both within the Church of England and in English politics. One of the primary difficulties, in addition to the diachronic change in definitions that the tract writer had noted at the time, is that the Puritans' agenda was never stated positively, but rather in terms of that what they opposed or rejected. What the Puritan movement stood *against* was the direction of the Reformation in England, and particularly the official policies of the state, which were regarded as having stopped the Reformation short of its proper goal.

---

[14] This was the claim of Charles and Katherine George in *The Protestant Mind of the English Reformation: 1570–1640* (Princeton, 1961), p. 6

[15] Patrick Collinson, *The Elizabethan Puritan Movement*, Berkeley and Los Angeles, 1967), p. 29. e.

For this reason, Collinson cautioned, "there is little point in constructing elaborate statements defining what in ontological terms Puritanism was and was not, when it was not a thing definable in itself, but only one half of a stressful relationship."[16]

However, it would be unfair to conclude that there was nothing that Puritans actually stood *for*, rather than against. Indeed, the individuals within the movement stood for a great many things, but the movement was, at any given time, most clearly united by that which it opposed. On this list of objectionable things were elaborate church ceremonies and any of the trappings of the Roman Catholic liturgy, but we cannot allow this to be the only entry, lest we mistakenly think, as did some opponents at the time, that Puritanism was mainly about external forms.

The doctrinal basis behind Puritan objections must likewise be recognized as separating Puritans from their opponents. Through the course of Collinson's treatment in *The Elizabethan Puritan Movement* Calvinism emerges as the most common doctrinal foundation of the movement.[17] This is not surprising, given the common scholarly recognition that the roots of Elizabethan Puritanism lie primarily with those exiles from the reign of the Catholic Queen Mary who took refuge in Geneva, Basel, Zurich, and other Reformed areas.[18] The Puritans did not look uncritically to Calvin and Geneva for guidance, but the very fact that their opponents appealed to Calvin in an attempt to quiet them and end the debate is evidence that the movement held Calvin in very high regard. We may note that it was only with reluctance that Thomas Cartwright, when confronted by Archbishop Whitgift, admitted that there were issues on which he and Calvin would disagree.[19] For this reason, it would be better to say that the Puritan movement was Reformed in theology, associating it with the branch of Protestantism of which Calvin was the most prominent figure, rather than suggest a specific allegiance to Calvin.[20]

To properly distinguish between set and subset, we must avoid conflating English Calvinism and the Puritan movement. There were plenty of English Calvinists who differed from their Puritan contemporaries in either emphasis or degree. Another reason why Whitgift quoted Calvin against Cartwright was that Whitgift was himself

---

[16]   Patrick Collinson, *The Birthpangs of Protestant England* (Houndmills, Basingstoke, 1988), p. 143.

[17]   Collinson, *The Elizabethan Puritan Movement*, pp. 36–7, 52–3.

[18]   On the Marian Exiles themselves see Christina H. Garrett, *The Marian Exiles* (Cambridge, 1938). Specifically in relationship to the rise of the Puritan movement in more recent scholarship see Collinson, *The Elizabethan Puritan Movement*, pp. 24, and 52–3 for concrete examples.

[19]   Collinson, *The Elizabethan Puritan Movement,* pp. 72, 104. We may also note that these differences were usually matters of casuistry – the practical application of the doctrines themselves – rather than differences of doctrine proper, as is the case of Cartwright on page 104.

[20]   Collinson is not concerned with such a subtle systematic distinction, but remains content with simply avoiding calling the Puritans "Calvinist." This distinction is important if we wish to keep Calvinism in perspective. Calvin gave Reformed theology practical expression in Geneva, and a degree of doctrinal definition which it had not achieved under previous theological leaders. In theological circles to this day, "Reformed theology" and "Calvinist theology" are treated as essentially coterminous when referring to the later sixteenth and beginning of the seventeenth centuries.

a Calvinist, as were the majority of English theologians at the time. But the Puritans were essentially concerned with getting Reformed theology right, since there were many issues of Reformed theology and practice where they felt that the institutional Church was failing. This brings us to another important aspect of the Puritan identity – its fundamental and vehement anti-Catholicism. Protestant though Elizabeth's Church of England was, it remained, for the Puritan in her reign (as it would in the reign of her successor), far too Catholic. To this extent, Trevelyan's old definition can still apply: Puritanism was "the religion of all those who wished either to purify the usage of the established Church from the taint of popery, or to worship separately by forms so purified."[21] In other words, Puritans were those English Calvinists who believed that, on any number of issues, the established Church had simply not gone far enough in rejecting Roman forms and religion and adopting the fullness of Reformed theology, and they took a stand on these issues because they felt them to be of critical importance. Beyond a clear basis in Reformed theology it is difficult to say what doctrines the Puritans positively held, and the details of what specifically was wrong and what was needed to correct it varied between one Puritan to the next.

The complexity of the Puritan question reflects the complexity of the era. We are dealing with a period of religious history that defies the systematization of simple categories. This is particularly true of the so-called *via media* of the old continuum, the "Protestants" mentioned in the poem, who were merely united by the two facts that they were not loyal to Rome, and that they did not share the concerns voiced by the Puritans regarding the established institutional religion. In Francis Bacon's lifetime the vast majority of the English population could be situated in this category, and, as the anonymous tract quoted earlier makes clear, to the Puritan or Roman Catholic observing from the outside it looked like a category of unhealthy compromise. However "compromise" can be a seriously misleading word. Those in this category were agreeing to conform to the Henrician Reformation, the Elizabethan Settlement, or the later official policies of the Crown. They were agreeing to tolerate one another, and they rejected any clearly partisan agendas which would divide the Church. Lack of adherence to partisan agendas should never be confused with being theologically "moderate" or with compromise at the level of personal belief. Rather than a compromise, the *via media* was, from the inside, an umbrella, covering those who had firm, though not homogenous, convictions, and who recognized that some latitude was necessary for good order in the realm. Edward Sackville, fourth Earl of Dorset, was one individual who represented the *via media* in the seventeenth century and Bishop Lancelot Andrewes was another. In considering them we can see the difficulties that the category itself presents for placing an individual within the context of the Tudor and Stuart religious milieu.

In an essay on Edward Sackville, David L. Smith has explored some of the difficulties that arise when we attempt to analyze an individual – especially one of the ruling and intellectual elite – according to the categories of the continuum.[22] During his lifetime, the Earl of Dorset "was called everything from a Puritan to a papist – and

---

[21]   As quoted in Dickens, *English Reformation*, p. 368.

[22]   David L. Smith, "Catholic, Anglican, or Puritan? Edward Sackville, Fourth Earl of Dorset, and the Ambiguities of Religion in Early Stuart England" in Donna

other things besides."[23] Dorset himself never kept a diary or made a convenient public announcement in which he said, definitively, what his own religious convictions were. His last will and testament, which might have given a clue to his convictions, is thoroughly ambiguous as well. The preamble to the will is a moderately Calvinist statement of faith, his executors were Roman Catholics, and his bequests were made to family and staff without consideration of religious affiliation. In the Star Chamber, Dorset had made a number of statements on behalf of religious toleration, but these, as Smith points out, do not reflect a "personal credo," but, rather, a "wider concern to preserve order" in the realm.[24] Dorset employed several domestic chaplains, but a distinction must be drawn between those whom he chose on account of their close personal contacts with his family and those whom he may have selected more freely. Without more specific personal evidence, Smith concludes that nothing definitive can be said about Dorset's personal beliefs beyond the recognition that he clearly represented what Peter Lake has called a "conformist cast of mind" in which one could avoid extremes and tolerate "a plurality of belief within a broad national church."[25] That he held such a general position of conformity does not mean that Sackville can be placed neatly on the continuum since, as Smith notes, "people of quite contrasting opinions could claim to be 'conformists' in early Stuart England."[26]

In contrast, there is no lack of information on the theological positions of Lancelot Andrewes. He left a wealth of sermons and personal devotional writings which clearly reveal a problem with the assumption that a conformist was necessarily interested in compromise, or in moderating the extremes of Catholicism and Puritanism. Andrewes might be said to represent his own extreme, which does not fit along the linear continuum at all. Nicholas Lossky has shown that, although Andrewes considered himself anything but a Papist, his theology was not typically Protestant either.[27] He was informed by his own reading of the Church Fathers, and his theology was shot through with ideas which, while common enough in Eastern Orthodoxy, are neither Protestant nor Catholic. He could have a great deal of sympathy for the Puritan focus on personal faith while insisting upon the necessity of Catholic liturgical forms and the Apostolic Succession. He was not governed by Western categories or systematic theology, and so he could also have a radical doctrine of free will which would have been condemned by Catholic and Puritan alike. Andrewes truly believed in, and advocated, the broad toleration which is the mark of the conformist. In his case it is clear that he was also the beneficiary of the broadly tolerant system which existed before his younger admirer William Laud became archbishop. Despite what the name implies, "conformity" in Andrewes' England was far from a lockstep affair, although it also had very real

---

B. Hamilton, and Richard Strier (eds), *Religion , Literature, and Politics in Post-Reformation England: 1540–1688* (Cambridge, 1996), pp. 115–37.

[23]   Ibid., p. 115.

[24]   Ibid., p. 118.

[25]   Ibid., pp. 127–8.

[26]   Ibid., p. 128.

[27]   Nicholas Lossky, *Lancelot Andrewes, the Preacher (1555–1626): The Origins of the Mystical Theology of the Church of England*, trans. Andrew Louth (Oxford, 1991).

boundaries. Conformity permitted Andrewes to freely consider the wide range of ideas and authorities which were to be found in Reformation England.

Both of these examples are relevant for our consideration of Francis Bacon. It was precisely Bacon's conformity which disturbed his mother, but, as with Dorset, merely saying that Bacon was a conformist does not tell us what he believed. Like his friend and mentor, Lancelot Andrewes, Bacon took advantage of the diverse range of ideas and texts which surrounded him and, also like Andrewes, he developed an idiosyncratic belief system from them. Dorset was thirty years younger than Bacon and lived in a far less tolerant time. As a result, Dorset needed to be far more reserved with his own opinions than Bacon or Andrewes. In considering any individual from this period we must remember that the habit of placing people on a continuum stretching from Catholicism to Puritanism misrepresents the diversity of opinion which was to be found. There was far more on the theological table of early modern England than Calvin's *Institutes* or the *Summae* of Aquinas, and the questions were much more complex than asking how these two systems should be balanced. Theologians, and intellectuals generally, had before them a smorgasbord of ideas and theological influences that would mix and blend as they were taken up or ignored, assimilated or rejected. The self-identified *via media*, far from being a "middle way" compromise between Catholic and Puritan, was an area of turbulent diversity of opinion from which only those who were divisive, particularly the Roman Catholics and the Puritans, were excluded. It was an area where "truth" could be sought rather than assumed. While Puritan and Papist each knew his respective truth, others, less clearly partisan but equally devoted, sought it out. Thinkers such as Andrewes and Bacon approached the problem with rigorous method and genuine reverence for the new edifice that they were constructing: a theology composed of truth, not polemic.

The Influences and Options Available in English Reformation Theology

So what was available for consideration by the theologians and intellectuals of Bacon's era? Nicholas Tyacke has emphasized the dominance of Calvinism throughout English society at this time, and this is a crucial first ingredient. The intellectual and theological world of Francis Bacon was a Calvinist world in which the non-Calvinists were a significant minority.[28] We may accept this statement with the same caveat that was applied to Puritanism earlier, namely that "Calvinist" is here used as a cover term for Reformed theology. However, Calvinism must not be allowed to overshadow the host of other trends and influences in Tudor and early Stuart religion. Tyacke also insists that non-Calvinists did exist. For them, Tyacke has coined the term "anti-Calvinist," reflecting the fact that they were in conscious tension with the dominant trend of English theology, and that these opponents to Calvinism pre-dated any movement in England which could legitimately be called "Arminian" as the Anti-Calvinists have often been characterized. Ultimately there was an "overthrow of Calvinism" in 1625, in which Arminianism itself could be said to have taken the field; but Tyacke's concern is important: we must not ignore the wide variety of thought present among the

---

[28]   Tyacke, *Anti-Calvinists*, pp. 1–2, 7.

non-Calvinist minority in England prior to 1625. The anti-Calvinist movement in England was gaining momentum long before the writings of Arminius were available.[29] There were also other alternatives. All of the main trends in continental Protestantism are represented in the English literature from this period.

Lutheranism, despite its failure as an option for unifying the Church of England, remained influential as a package of theological ideas throughout Bacon's lifetime. If the antipathy of Henry VIII made it less than prudent to adopt Lutheranism wholesale, the continuing influence of Lutheran thought, particularly on biblical exegesis and sacramental theology, can be traced in the writings of numerous individuals.[30] The Thirty-Nine Articles left plenty of room for all but the most dogmatic Lutherans to fit nicely into non-Calvinist corners of the Church of England.

Anabaptism is also commonly recognized as a component of the diverse English religious scene in this era. Exponents of the radical reformation of the continent, the Anabaptists, began to emigrate to England soon after Henry broke with Rome. Some of the first immigrants met with the same reaction they had been encountering on the continent and were promptly burned in St. Paul's churchyard. At no time were the Anabaptists accepted by the official Church of England, and they were the constant target of authorities in both the Church and the state, who resented their separatism as much as their radical doctrines.[31] Nevertheless, the English environment proved to be considerably more hospitable to Anabaptists than most areas of the continent, if only because the irenic policy of the Elizabethan Settlement precluded them from being rooted out wholesale as they were in genuinely Lutheran, Calvinist, or Catholic lands. In fact, England served as something of an incubator for Anabaptism. The movement continued underground, and various ideas of Anabaptist association floated through the English intellectual scene rather freely before England's own native Anabaptists eventually emerged to complicate the strife of the English Civil War.[32] It is as important to acknowledge that the movement provided the various parties of the Church of England with the unifying influence of having a common enemy as it is to consider their actual contributions to English thought.

Of course, apart from Calvinism, the most active and direct continental influence was that of the Roman Catholic Church. Reclaiming the island lost to the papacy was a special project of the Jesuit Order during this period, and its effect is not only to be measured in the number of actual converts to the Roman Church, such as Bacon's close friend, Sir Tobie Matthew. The continued presence of Roman Catholic voices challenging, and contributing to, the intellectual discourse of Reformation England led significant figures such as Lancelot Andrewes and William Laud to become concerned with questions of the continuity of the Church, the doctrine of Apostolic Succession,

---

[29]   On the "overthrow of Calvinism" see ibid., p. 8. On the ideological sources for this overthrow see p. 4, and chapters 1–4.

[30]   See Basil Hall, "The Early Rise and Gradual Decline of Lutheranism in England 1520–1660" in *Studies in Church History,* Subsidia ii (Woodbridge, 1979). See also the only other major work on the subject, Henry Eyster Jacobs, *The Lutheran Movement in England* (Philadelphia, 1890).

[31]   Dickens, *English Reformation*, pp. 262–68.

[32]   Ibid., p. 23. Note the influence that Dickens describes of Anabaptism upon the development of religious toleration – for example, on p. 379.

and, generally, the danger of throwing out the baby of Christian tradition with the bathwater of Roman abuses. Novel and logically consistent answers had to be found to the challenges of the Jesuits, such as "Where was the true Church before Luther?" and "How can those who do not repent of schism be saved?"[33] Thus the continued Catholic presence served both as a motor for intellectual activity and as an influence upon the development of "high-church" thinking within the Church of England.

Religion in England, and especially English Protestantism, cannot be understood simply in terms of continental developments. English theology was always marked by a uniquely English synthesis. Lollardy, in particular, had become thoroughly combined with English Protestantism by 1540.[34] Although after Wyclif's death it was less of a doctrinal position and more of what Alec Ryrie has aptly termed an "amorphous body of native heresies," Lollardy profoundly influenced the development and direction of English Protestant thought, and added to the complexity of the theological landscape of Reformation England.[35] By Bacon's time, Lollard doctrines could no longer be cleanly separated from broader Protestant discourse, and hence Lollardy *per se* is of very limited use in analyzing the theology of Bacon or his contemporaries. The incorporation of Lollardy into the theology of the English Reformation is an important reminder that the contours of English theology were never completely contiguous with continental theology. Long before the Reformation era, from Pelagianism through Lollardy, the island had earned a reputation for unique theological opinions. During the Reformation Wyclif and his movement became icons of English theological distinctiveness, and contributed to the justification for England's continuing to go in its own theological direction. If the lack of theological definition in Henry's break with Rome permitted theological diversity in England, the cultural icons of Wyclif and Lollardy had already nurtured a type of experimental thinking that only added to that diversity. In the Reformation England did not have to conform to prepackaged ideas any more than it had in the past.

Intellectual Trends: Patristics and Hebrew

Up to this point we have only been considering the elements of the early modern religious scene which could be called "factions" or "parties," and these are fairly

---

[33]   The first question is a common challenge of the Jesuits, particularly in Lutheran lands on the continent. It was apparently also found in England, as the tract *Luther's predecessours, or, An answere to the qvestion of the Papists : where was your church before Luther?* (London, 1624) attests. The second question is one raised by Tobie Matthew himself as the common theme of his tract, *Charitie Mistaken* (London, 1630).

[34]   Dickens (*English Reformation*, pp. 49–60) makes three important points in this regard: 1) Lollardy survived as a movement until the Reformation; 2) Lollardy prepared the way for Reformation doctrine in England; and 3) once the Reformation was underway, Lollardy was quickly supplanted as a movement by "Protestantism" generally. However, Lollard doctrines were not the same as those of continental Protestantism, and were, as Ryrie argues, as much of an obstacle to Lutheranism as an aid to its reception in England. See Ryrie, "Strange Death of Lutheran England," pp. 79–85.

[35]   Ryrie, "Strange Death of Lutheran England," p. 79.

typical of textbook accounts of the Reformation. However, we cannot understand these factions properly without paying attention to certain intellectual trends which are not so commonly considered. These trends fueled the Reformation on all sides and did much to contribute to the diversity of theological opinion in the early modern period, particularly in England. The first set of trends is linked to the recovery of ancient texts by the Renaissance humanists in the centuries prior to the Reformation. For the purpose of understanding Bacon's environment we will focus on two developments in particular: first, the recovery of the Church Fathers and other writings associated with Christian antiquity; and, second, the recovery of the Hebrew language and texts among Christians.

Along with other aspects of the Renaissance movement, the fourteenth century saw what Charles Stinger has called a "renaissance of patristic studies."[36] Thanks to the efforts of the humanists, the writings of the Christian theologians of the first seven centuries, both Latin and Greek, were gradually recovered and made public. Over time, as more and more ancient authorities came into circulation, this development dealt a serious blow to the method of medieval Scholastic theology. The internally consistent logical formulae of Scholasticism were at odds with the theological method, and often with the doctrines, of the ancient authorities.[37] Those differences posed no difficulty for many humanists, who saw little worth in the Scholastic method anyway. Erasmus, who stands at the apex of humanist theology, valued the Greek and Latin Fathers precisely because they demonstrated that medieval Scholasticism was a novelty. In his view, to return to the true *vetus theologia* – the original theology of Christianity – Scholasticism had to be abandoned.[38] By the time of the Reformation, interest in the expanding corpus of the Fathers had proceeded so far that the debates of the Reformation were saturated with continual citations of the Greek and Latin Fathers. Among Catholics and mainstream Protestants alike, it is difficult to find a scholarly theological work from the second half of the sixteenth or the seventeenth century that does not place tremendous weight upon the opinions of the early Church Fathers. The impact of the recovery of the Church Fathers extended far beyond theology to areas as diverse as cosmology, rhetoric, and civics. The humanist movement found the Fathers to be Christian authorities who themselves lived in classical antiquity, and were thus uniquely situated to justify and guide the humanists' use of pagan classical sources. The Greek Fathers reintroduced to the West a profoundly Platonic form of Christian theology which would have a significant influence on both the cosmology of humanists such as Pico, and, more generally, on discussions of the place of man in the universe. Yet the crisis which the recovery of the Fathers, and particularly the Greek Fathers, caused in theology was huge; it must be considered as central not only

---

[36] Charles L. Stinger, *Humanism and the Church Fathers: Ambrogio Traversari (1386–1439) and Christian Antiquity in the Italian Renaissance* (Albany, 1977), p. 83.

[37] Regarding the preceding decline of Patristic theology in the Middle Ages, especially of the Greek Fathers, see ibid., pp. 84–97.

[38] On Erasmus' concept of the importance of the *vetus theologia* see Istvan Bejczy' discussion in *Erasmus and the Middle Ages: The Historical Consciousness of a Christian Humanist* (Leiden, 2001), pp. 24–32, 104–5, 108–17, and 192–4.

to understanding developments in Catholic thought from the time, but also to the development of the Protestant Reformation.

The Catholics of Erasmus's day were saddled with reconciling the official theology of the Church as it had developed through the centuries with authorities which had fallen into disuse. There were also those who believed that the gulf between late medieval theology and the purer theology of the Early Church was simply too great to be reconciled, and, because of them, the unity of Western Christendom was lost. Protestantism, from its inception, shared with Erasmus the concern for recovering the *vetus theologia*. If the theology of the early centuries of Christianity could be identified, then all the errors of medieval Scholasticism, and of the papacy generally, would be clearly seen for the accretions that they were. The Church could then recover its original, and proper, theological emphasis and move forward from there. The Greek Fathers, who were problematic for the Catholic adherents of Scholastic theology, were a new arsenal for Protestants engaged in identifying Catholic error and supporting the break with Rome, though the Latin Fathers received equal attention; and the Patristic era also functioned as a unified authority for early Protestants. Closer to the fountainhead of Christianity, the Fathers were purer in their theology, and hence an important key to understanding when and where the Roman Church went wrong. But they were also studied positively as sources which gave clear precedent to Reformation theology and provided necessary insight into the original nature of Christianity. Thus the study of the Fathers was a central occupation of Protestants across Europe, and they were largely responsible for the development of Patristics as a discrete field of academic theology.[39]

The common Protestant concern for Patristic authority must qualify our understanding of the famous *sola scriptura* principle among early Protestants. For mainstream continental Protestants, such as the Lutherans and the Calvinists, the Bible was never the sole authority. It was the sole *absolute* authority, or the sole *infallible* authority. It was the authority by which other sources and authorities were to be measured and judged. Among Protestants of Bacon's era (with the exceptions of the more radical elements among the Puritans and of the Anabaptists generally) a theological argument was seldom considered complete without extensive reference to the opinions and decisions of the Early Church. Just how much authority the Fathers were allowed on any given issue was another matter. In Lutheran and Reformed lands, where there was an established doctrinal agenda for the Reformation, the Fathers were allowed to support the Lutheran or Reformed doctrines, but were rejected when they conflicted with the stated doctrines and confessions. For example, the Lutheran Martin Chemnitz makes use of both Basil and Epiphanius to support his defense of Lutheran doctrine against the Council of Trent, but both are rejected when their statements run foul of the Lutheran Confessions.[40]

---

[39] Johannes Quasten, *Patrology* (4 vols, Westminster, MD, 1984), vol. 1, p. 1. Quasten traces the earliest name of the field, *Patrology*, to the Lutheran Johann Gerhard, in 1653.

[40] Martin Chemnitz, *Examination of the Council of Trent*, trans. Fred Kramer (4 vols, St Louis, 1971). See vol. 1, p. 257 for the use of these Fathers for support, and pp. 287 and 267 for the rejection of the authority of Epiphanius and Basil respectively.

According to Chemnitz, these Fathers were already guilty of Romanist errors. Similarly, although Chrysostom is among Calvin's favorite Fathers to cite in support of the Reformed view of the Sacraments, he carefully distances himself from those passages in Chrysostom's writings which refer to the Lord's Supper as a "sacrifice."[41] What was pure in the Fathers and what was part of the imperfections already present in the early Church was determined by the established agendas of the Lutherans and the Reformed.[42]

In England, where the Reformation had no such clear agenda, the Fathers could be, and often were, given much more weight. Jean-Louis Quantin, in an article surveying the reception of the Fathers in seventeenth-century Anglican theology, portrays England as more focused upon Patristic theology than any location or group on the continent. This was also recognized at the time. Isaac Casaubon saw England as a refuge from both Catholicism and continental Protestantism, where he would be free to follow the theology of the ancient Church. Because of England's interest in Patristic theology and in the freedom to follow it, seventeenth-century London became a powerhouse for the production and publication of critical editions of the Church Fathers.[43] Lancelot Andrewes, for example, gave the Fathers a great deal more authority than was condoned by continental Protestants, and he often sided with the opinions of the Fathers over those of the Reformers. English theologians such as Andrewes were free to discuss the Apostolic Succession and the authority of tradition in ways that were theologically precluded on the continent. This would have a lasting impact on the doctrine and practices of the Anglican Church.

It is important to remember that the early modern West read the Greek Fathers with concerns and assumptions which were often far removed from the cultural context of the Fathers themselves, and hence Western thinkers often used the Fathers to draw conclusions which would have been foreign to the intent and understanding of the early Christian East. Nor were the orthodox Fathers recovered alone. A most notable example of an extra-Patristic text which shaped the theological landscape of Europe (along with many other intellectual developments of the Renaissance) is the *Corpus Hermeticum*, recovered and disseminated by Marsiglio Ficino in the late fifteenth century. Ficino believed that he had recovered part of the *prisca theologia*, the ancient or pristine theology, in this collection of philosophical and magical texts. The recovery of this supposedly ancient body of otherwise lost wisdom conformed to the idea of recovering lost learning that was driving the humanist movement in the Renaissance. There is also a kinship between Ficino's notion of a *prisca theologia* and Erasmus' concern for the recovery of the *vetus theologia*.

Certainly the *Corpus Hermeticum* itself was a controversial collection of texts. Not all intellectuals of the Renaissance would acknowledge Ficino's argument for its acceptability among Christians. However, the way in which it was supported

---

[41]   ICR, IV,18.11.

[42]   See Stinger, *Humanism and the Church Fathers*, p. 227 regarding the use of the Lutheran doctrine of Justification as a guideline for reading the Fathers among Melanchthon and his associates.

[43]   Jean-Louis Quantin, "The Fathers in Seventeenth Century Anglican Theology" in Irena Backus (ed.), *The Reception of the Church Fathers in the West: From the Carolingians to the Maurists* (Leiden, 1997), pp. 987–1008.

or rejected tells us much about the theological climate of early modern Europe. Proponents, such as Ficino, were bolstered in their assessment of the *Corpus* by the fact that the attitude of the Church Fathers, taken as a whole, was truly ambiguous toward the value of the *Corpus* and one could easily focus on the Patristic opinions that were positive. In addition, the Greek Fathers in particular held the same Platonic and neo-Platonic views of an immanent God, which are found in the *Corpus Hermeticum*. Opponents, such as Isaac Casaubon, focused on the inconsistencies between the *prisca theologia* and an established understanding of orthodox theology, focusing on negative statements by the Fathers. Between proponents and opponents there were those who weighed parts of the *Corpus* and embraced that which was found to be acceptable to the Christian worldview, even if the tradition itself was far from pure. The key is to recognize the role of theology, and especially Patristics, in measuring the value of these texts. All ideas in the seventeenth century were theological in their implications, if not in their very nature. That there was such a profound difference of opinion over whether the Hermetic texts were acceptable, and just how much of this tradition could be accepted, is a further testimony to the tremendous turbulence and diversity of early modern religious opinion.

Another important intellectual development in this era has received even less scholarly attention, namely the Christian rediscovery of the Hebrew language and the Hebrew Scriptures. This trend had profound effects on Christian theology and the Christian worldview in the early modern era. At stake was the meaning of two-thirds of the Christian Scriptures, and there were implications for natural philosophy as well as theology, particularly since it involved the creation accounts of the book of Genesis.

As with nearly every project designed to recover lost knowledge by a return *ad fontes*, the blossoming of the Christian interest in Hebrew in the early modern era had its beginning among the humanists of the Italian Renaissance, and in particular with Giannozzo Manetti and Giovanni Pico della Mirandola.[44] Other scholars soon followed, including Pico's disciple Johannes Reuchlin and the Dominican, Sanctes Pagninus. Pagninus undertook a new translation of the Bible from Greek and Hebrew in 1524, arguing that the ancient sources used by Jerome were admitted by Jerome himself to be unreliable and that the Church now had access to the resources to do a better translation. Despite the inherent criticism of the Church's official version of the Scriptures at a time when Protestants were busy criticizing the same, Pagninus' work received the blessing of Popes Leo X, Adrian VI, and Clement VII.[45] Papal approval, however, did not guarantee freedom from controversy.

Because of the need to rely upon Jewish sources, the study of Hebrew was by far the most controversial humanist undertaking to Roman Catholic theologians. The Christian scholars who undertook the study of Hebrew had to rely entirely upon the theologians of another religion in order to carry it off. The keepers of the Hebrew language were Jewish, and the Hebrew texts to which the Christian scholars turned were the product of generations of rabbinic transmission, and a prepackaged

---

[44] G. Lloyd Jones, *The Discovery of Hebrew in Tudor England: A Third Language* (Manchester, 1983), pp. 19–30; and Charles Trinkaus, *In Our Image and Likeness: Humanity and Divinity in Italian Humanist Thought* (2 vols, Chicago, 1970), vol. 2, pp. 578–600.

[45] Jones, *Discovery of Hebrew*, p. 40.

rabbinic interpretive tradition came along with them. Scholars who pursued the discipline were in the midst of a constant debate. Critics of the study such as Johannes Pfefferkorn, himself a converted Jew, roundly decried practitioners such as Reuchlin as the polluters of the true faith. While Pfefferkorn and his Dominican associates stoked fires with Hebrew texts, Reuchlin found his fellow humanists often reluctant to join him in an unqualified defense of the study of Hebrew.[46] Even among those who generally supported the study of Hebrew there was a concern over just how much a Christian scholar could trust or rely on rabbinic commentaries. Finally, there was a pervasive attitude among those not directly involved in the controversy that, regardless of the fascinating insights which Hebrew might offer, the study could only be of limited usefulness. For many, Hebrew had some value as a missionary and apologetic tool for the conversion of the Jews, but there could be no real point in redoing what Jerome had done correctly in the first place.[47]

Objections to the study of Hebrew were almost non-existent among the Protestants, who were not at all convinced that Jerome should have the last word on the Old Testament. Hebrew meshed well with the Protestant concern for getting the interpretation of the Scriptures right, and reforming the Church around the proper sense of the sacred text. As a discipline, Hebrew took off in Protestant centers of learning such as Wittenberg, where Reuchlin's precocious nephew, Philip Melanchthon, taught, and Basel, which Sebastian Münster turned into a center of Protestant Hebrew studies. Later, Oxford and Cambridge took up Hebrew as well. But the reliance on Talmudic interpretations, or rabbinic authority, was still regarded with much suspicion by Protestant scholars. Luther, for example, was convinced of the value of Hebrew, but was equally convinced that rabbinic exegesis had no place among Christians.[48] Calvin was cautiously ambivalent about the value of rabbinic sources.[49] On the other hand, Sebastian Münster's approach drew heavily upon the rabbis of all eras as authorities in Old Testament interpretation, and he was sharply criticized for this.[50]

As a result of the simultaneous rejection of Jerome and the rabbis as proper interpreters of the Hebrew text, there was a prevailing sense among the Protestants that the true Old Testament theology – the true meaning of the original text – was currently being recovered in its fullness. This belief in change and development in interpretation meant that there was a great deal of variation in Protestant Old Testament exegesis, which was lacking in that of Roman Catholics or Jews. The net effect of the early modern recovery of Hebrew may actually be regarded as a trend toward *less* uniformity in textual interpretation, rather than greater precision. More options were on the table than ever before, multiplied not only by Jewish exegesis, but also by the different theological agendas of the various Protestant groups.

---

[46]   Ibid., pp. 26–36.

[47]   This, for example, is the objection raised by Leonardo Bruni to Manetti. See Trinkaus, *In Our Image and Likeness*, vol. 2, pp. 578–600.

[48]   Jones, *Discovery of Hebrew*, pp. 56–66.

[49]   Ibid., pp. 76–7, 78–9.

[50]   Ibid., pp. 44–8.

Millennialism and the Belief in a Providential Age

In addition to these essentially intellectual developments which shaped early modern Christianity were other trends which were less precise though no less influential. One of these, which is of particular significance for a consideration of Bacon, is the cultural phenomenon of a widespread belief that humanity had just entered, or was on the threshold of, a special age decreed by divine providence. There have been many fine studies of Millenarianism and apocalypticism in the early modern period, but both of these phenomena are rightly conceived as subsets of something which is much more pervasive in early modern thought. Almost all of the literature dealing with millenarianism and apocalypticism as trends in the early modern period is concerned with popular movements and groups within a broader society. To recognize those groups and movements as distinct from the rest of early modern society, scholars make use of narrow and specific definitions of "millenarianism" or "apocalypticism," and confine themselves to very narrow aims. For example, in order to separate out those groups or movements which he wants to study, Howard Hotson uses a fairly typical example of a narrow definition: "Millenarianism, strictly defined, is the expectation that the vision described in the twentieth chapter of the Book of Revelation of a thousand-year period in which Satan is bound and the saints reign is a prophecy which will be fulfilled literally, on earth, and in the future."[51] Apocalypticism is usually more broadly defined by the omission of the strict interpretation of a period of one thousand years, but the emphasis on the Book of Revelation is still strong.[52]

Useful as they are, social and cultural histories of millenarian movements and trends of apocalyptic thought can obscure the fact that the specific movements and trends being considered are but elements of a much broader trend. For example, it is too easy for the modern reader of these histories to forget that the early modern critics of millenarian sects were themselves steeped in a culture of apocalyptic expectation and speculation, and held ideas which we would recognize as very similar to those which they were denouncing. Those who condemned a specific and rigid reading of the Apocalypse of John might well call upon other apocalyptic sections of the Old and New Testaments to do so: what was being rejected was not the idea of a special age, but a particular interpretation of what that age would entail. Throughout early modern Europe there was a widespread belief that a special age had or would soon

---

[51] Howard Hotson, "The Historiographical Origins of Calvinist Millenarianism" in Bruce Gordon, (ed.), *Protestant History and Identity in Sixteenth-Century Europe* (2 vols, Aldershot, 1996), vol. 2, p. 160.

[52] Richard Bauckham, for example, begins with the broadest possible definition of "apocalyptic," but soon moves to in-depth discussions of Revelation and the literal descent of the New Jerusalem, etc. See Richard Bauckham, *Tudor Apocalypse: Sixteenth Century Apocalypticism, Millenarianism, and the English Reformation* (Oxford, 1977), pp. 14–16. See also Katharine R. Firth, *The Apocalyptic Tradition in Reformation Britain: 1530–1645* (Oxford, 1979); and Paul Christianson, *Reformers and Babylon: English Apocalyptic Visions from the Reformation to the Eve of the Civil War* (Toronto, 1978). While other biblical texts are also considered "apocalyptic" (Daniel, 2 Thessalonians, etc.) the trends being described involve a Johanine reading of these texts as well.

come upon them in which momentous changes, wrought by the hand of God, would transform the world, and that such an age was foretold in the Scriptures.

This belief in a glorious providential age was especially pervasive among Protestants, who saw the Reformation as the threshold of just such an age.[53] It was also a concept that was informed by the tremendous political changes which were taking place in early modern Europe. The emerging sense of national identity in Europe was enmeshed with ideas of divine favor or disfavor, and the common belief that God was raising up a particular chosen people for His special work. This was an age when the providential hand of God was beginning to operate within the bounds of nations. For the Spanish, the sailing of the Armada was "God's obvious design," while the English saw the obvious design of God in its failure.[54] In Bacon's own writing, as well as that of his followers, there can be found the conviction that Britain, her king, and her people, were set aside by God for a particular glorious destiny.

Bacon's Break with the Godly

When Francis Bacon left home there was indeed a wide world of alternatives to his Puritan upbringing, and Francis left home for the first time at an early age. In 1573, at the age of twelve, he joined his brother Anthony at Cambridge. The two boys were housed and tutored by none other than John Whitgift, whose influence over Francis would later be lamented by Anne in her letter to Anthony of 1592. From his later writings it is clear that Francis Bacon was a creature of the religious landscape we have just toured. He was intimately acquainted with the Church Fathers and had chosen his favorites among them. This would not have necessarily disturbed Anne, who read the Fathers in Latin and Greek herself and whose sister, Margaret, had made a translation of St Basil's sermon on Deuteronomy 15.[55] What Bacon himself got out of the Fathers, however, was far from reinforcement for his mother's beliefs, for he shared the widespread belief that his society was on the cusp of an important age decreed by providence, but his conception of that was unique, as we shall see. Although there is no evidence that Bacon was personally qualified in the Hebrew language, he became well acquainted with those who were, and the flexibility of

---

[53] The images used in Reformation polemic make the apocalyptic significance of the Reformation vividly clear. See Charles Scribner, *For the Sake of Simple Folk: Popular Propaganda for the German Reformation* (Cambridge, 1981). Consider also Jaroslav Pelikan's discussion, "Some uses of Apocalypse in the Magisterial Reformers" in C.A. Patrides and Joseph Wittrich (eds), *The Apocalypse in English Renaissance Thought and Literature* (Ithaca, NY, 1984), pp. 74–88. For Catholics, who interpreted the Reformation differently, the sense of an imminent change was neither so pervasive nor so positive, though among Catholics as well there was a widespread expectation that God would soon end the schisms and restore the unity of Christendom. Also consider the more overt forms of Catholic millenarianism discussed in Karl A. Kottman (ed.), *Catholic Millenarianism: From Savonarolla to the Abbé Grégoire* (Dordrecht, 2001).

[54] After the death of Mary Queen of Scots, Philip's ambassador in Paris wrote that it was "God's obvious design" to use him and give him the kingdoms of England and Scotland. See P. Gallagher and D.W. Cruikshank (eds), *God's Obvious Design* (London, 1988), pp. vii, 167.

[55] Jardine and Stewart, *Hostage to Fortune*, p. 25.

interpretation which marked the reading of the Old Testament at the time permitted him to reconsider both prophecy and the creation narrative of Genesis in his own way. Lady Anne Bacon's concern for her younger son was well founded. As an adult, Francis had left his Puritan heritage behind, and, three years prior to Anne's letter to Anthony, Francis had made his position known in a public statement.

In 1589 Francis Bacon circulated a tract in manuscript form entitled *An Advertisement Touching the Controversies of the Church of England*, which weighed in on the Marprelate controversy, a tract war begun by the vitriolic Puritan who went by the pseudonym "Martin Marprelate." James Spedding has aptly characterized this exchange as "that disgraceful pamphlet war which raged so furiously in 1588 and 1589 between the revilers of the bishops on the one side, and the revilers of the Puritans on the other, and in which the appeal was made by both parties to the basest passions and prejudices of the vulgar."[56] The controversy between the two sides began years earlier when Bacon was a student at Cambridge, and his close connection with both sides gave him a unique perspective.

Bacon's *Advertisement* has never been subjected to a close reading in order to understand the theological points made there. Most interpreters have characterized it as a simple call for toleration and compromise, designed to ensure that he was properly situated politically on issues of religion, and having little theological substance.[57] Bacon does counsel moderation and toleration in this tract, but he does so in a way which goes far beyond mere political propriety. The tract reveals a keen grasp of Church history and is theologically nuanced. Whether at Whitgift's hands or elsewhere, Bacon had been trained well. It would have been foolish to wade into such a vicious controversy without having the skill to navigate the subtleties of theology. The key players on both sides were skilled in the field. Even when discussing "matters of indifference" one had to take a theological stand in the sixteenth and seventeenth centuries, for the question of *which* issues were actually indifferent was theologically volatile wherever it arose. To advise moderation on theological matters required a well-informed theological position in an era when theology was so charged, but Bacon does more than advise moderation. He makes significant theological points of his own, which demonstrate that the breadth of the *via media* had allowed him to develop in his own way.

Bacon's tract begins with a lament that such a controversy should have begun among Protestants. This fight was over things which were not part of the essential "mysteries of faith" which precipitated the Christological controversies resolved by the Church Fathers in the early Councils. Rather, the two sides are divided over matters which, for the most part, are matters of "indifference" pertaining to ceremonies and the "extern[al] policy and government of the church." He calls upon both sides to adopt the more conciliatory approach of the Apostles and early Church

---

[56] WFB, vol. VIII, p. 72.

[57] This is the conclusion of Perez Zagorin in *Francis Bacon* (Princeton, 1998), pp. 6–7; also Julian Martin, *Francis Bacon, the State, and the Reform of Natural Philosophy* (Cambridge, 1992), p. 38. See also Jardine and Stewart, *Hostage to Fortune*, p. 125, who suggest that this was a mere act of self-fashioning which went against his own personal Puritanism, which he retained.

Fathers, "which was, in the like and greater cases, not to enter into assertions and positions, but to deliver counsels and advices" and if this were done, the English Church "should need no other remedy at all."[58] The most lamentable error of each side lay, according to Bacon, in their insistence that the other side was ungodly. Like the Earl of Dorset, Bacon was a conformist primarily interested in quelling controversy. But he was far from suggesting that the truth was to be found in the middle. Although the rhetorical tone of Bacon's *Advertisement* is mediating, placing the blame for the controversy squarely upon both sides, it also gives evidence of Bacon's personal opinions and biases in the matter, particularly when we consider *what* Bacon finds blameworthy in either side.

The bishops, and those who side with them, are reprimanded for their intractability, their heavy-handedness, and their failure to listen to the other side. In short, this party is to blame for the harshly dismissive *way* in which they handle criticism and objections. Bishop Thomas Cooper, the first to respond to "Martin Marprelate," did so admirably, according to Bacon, not sinking to "Marprelate's" level, but by reverently responding to the issues raised, rather than to the anonymous person and his language. Others did not follow the example of Bishop Cooper and are truly guilty of making the matter into a major controversy; for, as Bacon quotes, "he that replieth multiplieth."[59] Among the party of the bishops, Bacon noted, "there is not an indifferent hand carried toward these pamphlets as they deserve."[60] While they began well, they had since become increasingly firm on all matters, denying that there was anything in the Church of England that could benefit from change or reform. The bishops themselves cannot avoid blame "in standing so precisely upon altering nothing," and they are certainly guilty of "unbrotherly proceeding" in charging the opposition "as though *they denied tribute to Caesar*, and withdrew from the civil magistrate the obedience which they have ever performed and taught." The Puritans, regardless of what the bishops may think of their doctrine, were not to be confused with radicals such as the Family of Love; and the bishops had all too eagerly believed every accusation against the Puritans by which they could be unfairly condemned.[61]

The Puritans were subject to a much weightier censure in Bacon's *Advertisement*. They were also guilty of "unbrotherly proceeding" in their harsh attacks on the bishops, who were often godly and "men of great virtues."[62] They had also gone to an extreme in insisting that matters which should be indifferent were not, and then hiding behind the "honorable names of sincerity, reformation and discipline ... so as contentions and evil zeals are not to be touched, except these holy things be thought first to be violated." While claiming to be for "reform," the Puritans were in fact perpetuating controversies merely for the sake of change. Quoting an unnamed "father." Bacon concluded: "They seek to go forward still, not to perfection, but to

---

58   WFB, vol. VIII, p. 75.

59   Ibid., p. 77.

60   Ibid., p. 78.

61   Ibid., pp. 87–9.

62   Ibid., p. 81.

change."[63] The Puritans were genuinely responsible for the schism in the Church even as Whitgift claimed, for "they refuse to communicate with us, reputing us to have no church."[64] Bacon's conformity is evident in the use of the first-person plural.

There was, according to Bacon, evidence that the Puritans had gone to a dangerous extreme in many of their doctrines, and he associated them with certain heretics. He noted how those most firmly opposed to Arianism in the early Church came to the equally heretical position of Sabellius.[65] Even so, the Puritans, by the vehemence of their opposition to all things which smacked of Roman Catholicism, had gone to the opposite extreme and begun to discard the good with the bad.[66] Thus, like those of old who became heretics by virtue of their zeal against heresy, the Puritans had chosen a questionable path in their objections to the Church when they claimed that reformation had not gone far enough:

> As in affection they challenge the said virtues of zeal and the rest, so in knowledge they attribute to themselves *light* and *perfection*. They say, the Church of England in King Edward's time and in the beginning of her Majesty's reign, was but in the cradle; and the bishops in those times did somewhat for daybreak, but that maturity and fullness of light proceeded from themselves. So Sabinus, Bishop of Heraclea, a Macedonian, said that the fathers in the Council of Nice [the first ecumenical council of Nicea] *were but infants and ignorant men; and that the church was not so to persist in their decrees as to refuse that further ripeness of knowledge which the time had revealed* ... so do they censure men truly and godly wise (who see into the vanity of their assertions) by the name of politiques; saying that their wisdom is but carnal and savouring of man's brain.[67]

As a function of the Puritan love for simplicity, which comprehended reliance on Scriptural precedent for all things and the preaching of the Word in sermons as the central mark of the true Church, they had developed a dangerously narrow view of Christianity:

> But most of all is to be suspected, as a seed of further inconvenience, their manner of handling the Scriptures; for whilst they seek express Scripture for everything' and that they have (in manner) deprived themselves and the church of a special help and support by embasing the authority of the fathers; they resort to naked examples, conceited inferences, and forced allusions such as do mine into all certainty of religion. Another extremity is the excessive magnifying of that which, though it be a principal and most holy institution, yet hath it limits as all things else have. We see wheresoever (in manner) they find in the Scriptures the *word* spoken of, they expound it of preaching. They have made it almost the essence of the sacrament of the supper, to have a sermon precedent. They have (in sort) annihilated the use of liturgies, and forms of divine service, although the house of God be denominated of the principal, *domus orationis*, a house of prayer, and not a house of preaching. As for the life of the good monks and the hermits of the primitive church, I know they will condemn a man as half a Papist, if he should maintain them as other than profane, because they heard no sermons. In the meantime, what preaching is, and who may be said to preach, they make no question. But as far as I see, every man that

---

[63] Ibid., p. 83.
[64] Ibid., p. 86.
[65] Ibid., p. 83.
[66] Ibid., p. 84.
[67] Ibid., p. 91.

presumeth to speak in chair is accounted a preacher. But I am assured that not a few that call hotly for a preaching ministry deserve to be of the first themselves to be expelled.[68]

Bacon's particular equation of the errors of Puritans with ancient heresy suggests that he was aware of, and utilizing, the historic definition of "heresy": from the Greek, αἱρέομαι, "to choose." Heretics were those who "chose" to separate themselves from the Church, thus dividing the Church, or who "chose" certain aspects of the faith to emphasize at the expense of others. Bacon's discussion concludes as it began, with a plea that both sides of the controversy rein in their invective before the situation becomes worse.

It must be remembered that Bacon's argument in the context of the Marprelate controversy was directed principally against the more extreme Puritans as they were to be found in the late 1580s. Many who were themselves of a more Reformed frame of mind and would even be regarded as Puritans in other contexts would have agreed with his basic tolerance. However, in the course of his discussion, Bacon went far beyond a mere call for peace, and expressed sympathies for a great many ideas which would not sit well with even more moderate Puritans. He wrote a number of things which would come to be regarded as "high church" when the term came into general use in the eighteenth and nineteenth centuries. While the exact *form* of ceremonies might be open to debate, the importance of ceremonies, or the ordered prayer of the Liturgy, was not, and the efficacy of the Lord's Supper was neither to be eclipsed by, nor made conditional upon, sound preaching. The authority of the Scriptures, though ultimate, was not exclusive, for the Church Fathers would thus be denied to the Church. Monks and hermits – at least those of the primitive Church – were mentioned with reverence, without any suggestion that the monastic calling was itself dangerous or flawed (a point that will resurface in Bacon's later writing). On the whole, this suggests that Bacon held a very high view of many things that are not explicitly sanctioned by the Scriptures, at least as the Puritans read them. And, for Bacon, Puritanism itself, by insisting on its own rectitude and necessity, was always in danger of genuine heresy. Although Bacon's censure of the bishops may have been moderated by his political situation, there can be no question that he clearly distanced himself from the Puritans in this tract. Significantly, his objections to the Puritans were theological, whereas his objections to the positions of the bishops centered on issues of casuistry and behavior.

Were the *Advertisement* a work written in isolation a claim still could be made that the positions expressed in it are not necessarily those of Bacon himself. But other works written during the next decade add greatly to our understanding of the development of Bacon's religious thought prior to the publication of his works pertaining to the Great Instauration. The *Advertisement* is indicative of more profound changes in Bacon's beliefs as he reconsidered many of his society's common theological assumptions. At the heart of his re-evaluation were questions about the nature of the relationship between God and creation, and the special place of human beings in the order of things.

---

[68]   Ibid., p. 93.

After 1594 Lady Anne Bacon's mental and physical health began to decline. We know very little of what may have passed between mother and son during the later years of her life. In 1600 Francis made a passing reference to Anne's "worn" health in a letter. It is the last reference to his mother in his writings prior to an expression of grief at her death in 1610.[69] In 1602 Lady Bacon signed over a number of manors to Francis as her heir. This may have been nothing more than a legal formality since her only other son, Anthony, had died several months before.[70] The sense of estrangement between Anne and Francis in these years is hard to avoid, and it adds a melancholy chord to his life. Francis's turn away from Puritanism disappointed his mother, but it also proved that he was her heir in more than her estates. From her he had received the education, the piety, and the passion for a correct understanding of the faith which would shape everything that he wrote.

---

[69] Jardine and Stewart, *Hostage to Fortune*, p. 321.
[70] Ibid., p. 253.

Chapter 2

Bacon's Turn toward the Ancient Faith

The Formative Years

Near the end of his life, in 1625, Francis Bacon wrote to Father Fulgentino, a Venetian priest, that he made his first attempt at writing down his thoughts on the reform of natural philosophy – what would become his "Great Instauration" – at the age of twenty-four, when he was in London, in 1585.[1] By his own confession to Lord Burghley, in a letter of 1592, he was at that time obsessed with his plans for the reform of learning, and was eagerly seeking a position which would allow him to direct his attention completely to natural philosophy.[2] The 1592 letter is the earliest surviving description of what the Instauration would involve. His description is detailed, and suggests that the basic idea of the need for a reform of learning had been well thought out during the preceding years. The years of the late 1580s and the 1590s were, as Markku Peltonnen has described them, a crucial period of Bacon's "intellectual gestation."[3]

During these same formative years Bacon was wrestling with, and breaking away from, the key features of the Reformed theology which dominated Elizabeth's realm. He was leaving behind not only Puritanism but also Whitgift's Calvinism, and turning toward the theology of specific Fathers of the ancient Church. The two pursuits of natural philosophy and theology were singular for him, as they had been for many Christians throughout the preceding centuries, for the Author of the Scriptures, which were the source of the true faith, was also the Author of the Book of Nature, which was Bacon's primary text for natural philosophy. Bacon's turn toward Patristic theology was more than a personal quest to come to the proper understanding of the faith. It was a turn toward the theology which supported the Great Instauration, and which is found in the theological discussions embedded in Bacon's later writings pertaining to natural philosophy.

As Lady Anne Bacon had noted in her letter to Anthony, there was no shortage of those who might encourage a young man to abandon the higher principles of Puritanism. Archbishop Whitgift was merely the one who had the opportunity to influence both of her sons: Francis would associate with quite a few others. Once he became established in the early seventeenth century, Bacon hand-selected personal chaplains, William Lewis and William Rawley, who were notable conformists with

---

[1] WFB, vol. XIV, pp. 375–7.
[2] WFB, vol. VIII, pp. 108–9.
[3] Markku Peltonnen (ed.), *The Cambridge Companion to Bacon* (Cambridge, 1996), pp. 4–5.

what would one day be called "high-church" tendencies.[4] The man who would, after 1601, be his closest friend, Tobie Matthew, would take orders as a Roman Catholic priest. Of those with whom Bacon associated as members of his literary circle, those who advised him on his writings, the only one who might genuinely be associated with puritanism was Thomas Bodley. Bodley and Bacon, however, had a very short-lived working relationship which ended when Bodley accused Bacon of grave errors which opposed the Scriptures.[5] Most of those who would assist Bacon with his Instauration writings had yet to make his acquaintance in the last decade of the sixteenth century. From this formative period the most significant figure was Lancelot Andrewes. There are good reasons to believe that there was an extensive exchange of ideas between Bacon and Andrewes, not the least of which is their similarity in matters of theology.

Bacon and Andrewes

The friendship between Bacon and Andrewes may be traced to sometime after 1589, when Andrewes moved to London and assumed the triple position of rector at St Giles' Cripplegate, and prebend at both Southwell and St Paul's. The friendship was well established by 1592, when Bacon sent Andrewes a cordial letter inviting him to join him in the country to avoid the plague that which was said to be in London that year.[6] However, Andrewes declined, being too dedicated to his flock to leave his post. Thus, within a short time of his composition of the *Advertisement Touching the Controversies of the Church of England*, Francis Bacon had developed a friendship with a notable "anti-Calvinist" and celebrated scholar of Patristic theology. More than a casual friend, Andrewes was also one of Bacon's editors when Bacon began publishing his works on the reform of natural philosophy. In this capacity Andrewes was greatly respected by Bacon.

A letter from Bacon to Lancelot Andrewes of 1609 (which accompanied a later draft of Bacon's *Cogitata et Visa*) is particularly helpful for understanding their friendship.[7] Bacon greeted Andrewes as a friend of many years, and in conclusion stated explicitly that, if he had been able, he would have come to him personally,

---

[4]   On Lewis see the entry in the old DNB, vol. 11, p. 1078 where he was described as a "zealous member of the high-church party." Both chaplains later served as chaplains to Charles I, and Rawley also served in this capacity to Charles II. The fortunes of both were lost in the civil war and restored by Charles II. William Rawley will be discussed at greater length in Chapter 6.

[5]   The literary circle will be discussed in Chapter 6. For more information on Bacon's literary circle, see Steven Paul Matthews, "Apocalypse and Experiment: The Theological Assumptions and Religious Motivations of Francis Bacon's Instauration" (unpublished dissertation, Gainesville: University of Florida, 2004).

[6]   The letter is found in WFB, vol. VIII, p. 117.

[7]   WFB, vol. XI, p. 141. Andrewes had just finished writing his *Responsio ad Bellarmine*, which had been commissioned by the king. Not being a controversialist, this had proven something of a distasteful task for him (p. 140). Bacon mentioned in the beginning of the letter, that a diversion into natural philosophy might now be welcomed by him, suggesting that he had been waiting until the prior task was off of Andrewes' table.

rather than send a letter, but both men were busy, and he was hastening to his house in the country. The letter contained a special request:

> I send not your Lordship too much, lest it may glut you. Now let me tell you what my desire is. If your Lordship be so good now, as when you were the good Dean of Westminster, my request to you is, that not by pricks, but by notes, you would mark unto me whatsoever shall seem unto you either not current in style, or harsh to credit and opinion, or inconvenient to the person of the writer; ... And though for the matter itself my judgment be in some things fixed, and not accessible by any man's judgment that goeth not my way: yet even in those things, the admonition of a friend may make me express myself diversely.[8]

James Spedding points out that the service which Dr Andrewes was to perform for his friend was not an uncommon one. Andrewes had read and edited Bacon's drafts of *Of the Proficiencie and Advancement of Learning* of 1605, and had evidently done a thorough job, for in sending the finished product to Tobie Matthew, his other main editor, Bacon wrote: "I thought it a small adventure to send you a copy, who have more right to it than any man, except Bishop Andrewes, who was my inquisitor."[9] Andrewes, then, was a common resource for Bacon as an editor and advisor on his projects, and he took to the work with a certain thoroughness and zeal. Though Bacon was not about to alter his main argument for anyone, he was entirely willing to change his presentation on Andrewes' advice. Furthermore, he had found Andrewes' advice particularly valuable and trusted him to ensure that nothing that he had written would offend, or place him in an "inconvenient" position. The friendship between the two was evidently a very candid one, in which such business was usually conducted face-to-face. Bacon, ever-cautious, was not one to expose his ideas before they were safe for public eyes, and Andrewes possessed not only the necessary level of erudition, but also the safety of being a long-time, close friend. Neither man could be said, in 1609, to represent the mainstream of theology in England, but Andrewes was by this time a bishop who was highly respected on all sides for his piety and irenicism. Bacon could not have a better advisor as he prepared to publish works which many of the mainstream Calvinist divines would find questionable.

Andrewes has come to be known as one of the "fathers" of Anglo-Catholicism. His theology, and that of his followers, is marked by a concern for the authority of tradition, an emphasis on ritual prayer (liturgy) with a concomitant de-emphasis on the value of preaching in the Sunday service, and a concern for maintaining the Apostolic Succession. Although Andrewes has been commonly associated with Arminianism on the basis of his firm rejection of Calvinist predestination, it should be noted that, in these other points, the genuine Arminians remained true to their Reformed roots while Andrewes expressly rejected the Reformed position. Representatives of Dutch Arminiansim, such as Grotius, were tolerant of differences on such issues, admitting them as *adiaphora* or "matters of indifference." For this reason Grotius could claim that Andrewes was an ally and a sympathizer. Andrewes, however, never admitted that differences on these points were insignificant. Therefore, although he advocated tremendous breadth and toleration in the Church of England as a practical matter,

---

[8] WFB, vol. XI, p. 141.
[9] WFB, vol. X, p. 256.

he did not compromise in his own theology. When the early Arminians in England sought to claim him for their number, he vigorously dissociated himself from them.[10] The sources of Lancelot Andrewes' theology were not to be found in any contemporary debates or partisan movements.

In the most nuanced study of Andrewes' theology to date, Nicholas Lossky established, through a careful reading of Andrewes' *LXVI Sermons*, that Andrewes' reverence for the Church Fathers, and Christian antiquity generally, led him to develop his own theological system which, though far from unique, was fundamentally Patristic, with a decidedly Eastern turn.[11] Our image of Andrewes should be dominated by his affinity for tradition and the Fathers, rather than an association with any of the contemporary flavors of Protestantism. Andrewes' theology was not standard for his day, but it was founded in authorities which, by virtue of their antiquity and their accepted orthodoxy, could not be rejected either.

Among other Patristic doctrines, Lossky demonstrated that Andrewes adhered to the Eastern Patristic understanding of salvation as *theosis*, or deification. In deification the doctrine of salvation is not limited to the action of Christ making atonement for sin in the crucifixion and the subsequent guarantee of life provided by the resurrection. Instead, the incarnation, the act by which God became a human being, is regarded as the pivotal point. In taking on human flesh and thereby uniting himself to all of humanity, God determined the nature of human salvation. It is through this "hypostatic union" that man sees his ultimate destiny, which is to take on the nature of God and become "deified." The cross and resurrection are a central part of the incarnation package and are necessary to make salvation possible, but the emphasis, even when considering the crucifixion, is on the union of humanity with God. As it was summarized in the Christian East from the second century onward, "God became man so that man might become God." A typical passage from Andrewes to this effect is found in his sermon on the Feast of the Nativity for 1609:

> To have made Him a body and taken it upon him for a time till He had performed His embassage, and then laid it off again, that had been much; but so to be made as once made and ever made; so to take it as never to lay it off more, but continue so still, γένεσθαι, 'it to become His very nature;' so to be made is to make the union full. And to make the union with us full, He was content not to be sent alone but to be made; and that γένεσθαι, 'to be made so as never unmade more.' Our manhood becoming His nature, no less than the Godhead itself. This is *Filium Factum* indeed.[12]

As God and humanity were united in the person of Christ, so now Christians, who are united with Christ through baptism, have themselves taken on the nature of God and are daily growing more "godlike" in their lives. This was the source and nature

---

[10] Nicholas Tyacke, *Anti-Calvinists: The Rise of English Arminianism, 1590–1640* (Oxford, 1987), p. 91. While there were good political reasons for doing so, as Tyacke notes, the differences in theology should not be minimized.

[11] Nicholas Lossky, *Lancelot Andrewes, the Preacher (1555–1626): The Origins of the Mystical Theology of the Church of England*, trans. Andrew Louth (Oxford, 1991).

[12] Lancelot Andrewes, *Works of Lancelot Andrewes*, ed. John Parkinson (11 vols, Oxford, 1854), vol. 5, p. 52.

of piety for Andrewes, and it was manifest in his own life. Andrewes never married. His personal piety led him to live a life of near-monastic asceticism and seclusion, with his days dominated by academic study and prayer, and he advocated in his sermons some level of ascetic discipline for all Christians.[13]

As well as the doctrine of deification, Andrewes preached the correlative doctrine that, as a result of God taking on human nature, humans would be glorified above the angels, and through union with Christ, human nature would "enter upon a state that no man had ever known before, neither the righteous ones of the Old Testament, nor even Adam before the fall."[14] In a 1615 sermon on the incarnation Andrewes used the eighth Psalm to expound the state of humankind, which, as a result of the incarnation, excelled that of the angels:

> "Lord, what is man," either Adam or Abraham, "that Thou shouldest be thus mindful of him, or the seed, or sons of either, that Thou shouldest make this do about him!" The case is here far otherwise – far more worth our consideration. There, "Thou hast made him a little lower" [than the angels]; here, "Thou hast made him a great deal higher than the Angels." For they, this day first, and ever since, daily have and do adore our nature in the personal union with the Deity.[15]

This doctrine was not unfamiliar in early modern Europe. It may be found in the *Oration on the Dignity of Man* of Giovanni Pico della Mirandola, among other places, but, wherever it is found, it is evidence of a particular Patristic influence.

Andrewes was also decidedly non-Western in his understanding of original sin. Although this has led to discussions of whether he should be classified as a "Pelagian" or "Semi-Pelagian," Lossky's connection of Andrewes to Eastern theology makes such Western classifications inappropriate. Andrewes, in turning to the example of the Eastern Church, avoided (or disregarded) both sides of the rift between Augustine and Pelagius. As a consequence, the sense of the *completeness* of the corruption of the human will which dominates Western discussion since Augustine is simply absent in Andrewes' writing.[16]

It is also noteworthy that the idea of *recapitulation*, as it is found in St Irenaeus of Lyons, permeated Andrewes' preaching. Although recapitulation is associated with the understanding of salvation as deification, it has implications which extend beyond the salvation of humankind to include the restoration of all creation as a result of the incarnation and human salvation. According to recapitulation, salvation involved nothing less than the complete restoration of Adam's relationship to God and creation in paradise. The relationship would be restored in a more perfect way, transforming both humanity and the rest of creation, for now God Himself had united with creation through the incarnation, and "sums up all things in Himself."[17]

---

[13] Maurice F. Reidy, S.J., *Lancelot Andrewes: Jacobean Court Preacher* (Chicago, 1955), pp. v, vi, 4, and 152–84.

[14] Lossky, *Lancelot Andrewes*, p. 49.

[15] Andrewes, *Works of Lancelot Andrewes*, vol. 1, p. 14.

[16] Lossky, *Lancelot Andrewes*, pp. 170–73.

[17] See especially Andrewes' extended discussion in his Christmas sermons of 1622 and 1623. Most of the content of these sermons reflects Irenaeus' writing specifically, though he does

There is very little that Andrewes said or believed which could not be traced to its roots in the writings of the first seven centuries of the early Church. Lossky's presentation of Andrewes' theology is based on an analysis of his *LXVI Sermons*, but the Patristic themes which he has identified are found throughout Andrewes' writings, from his widely known manual for personal devotions, the *Preces Privatae*, to his early lectures at St Paul's and St Giles, Cripplegate. These early lectures were preached early in his friendship with Bacon, in 1591 and the following years, and are preserved in the posthumous publication, Αποσπασματια *Sacra*. The doctrines of Andrewes that we have just considered, relating to the way of salvation, have special significance for the theological statements of Francis Bacon: during the last decade of the sixteenth century Bacon was working his way toward very similar positions.

The *Meditationes Sacrae* and Bacon's turn away from Calvinism

Throughout his life Bacon would use Calvinist language and terminology to wrestle with Calvinist questions. This may be understood as evidence of the degree to which English theology was dominated by the constructs and formulae of the *Institutes*, but it certainly also supports the notion that Bacon's own Calvinist heritage had a lasting influence on his thinking. A careful reading is required to recognize what Bacon is doing in his use of language. The mere use of Calvinist language does not preclude serious disagreement with Calvinism. As with many others at the close of the sixteenth and dawn of the seventeenth centuries, Bacon became dissatisfied with Calvin's answers to many questions of the faith. One marker along the path of Bacon's departure from Calvin's influence is the *Meditationes Sacrae*, which he first published in 1597. Benjamin Milner has noted that although the *Meditationes* generally deal with fairly common themes from the Protestant Reformation, in dealing with the particular topics of "Atheism" and "Heresy," Bacon has adopted a distinctly Calvinist approach.[18] Yet, if the approach is Calvinist, the conclusions are not. Bacon does not merely reproduce Calvin's arguments, even if he borrows their language and outline, and in the case of his meditation on heresy he goes out of his way to distance himself from one of Calvin's more controversial doctrines.

It is clear from all of Bacon's *Meditations* that, when pursuing questions of the faith and theology, the significance of these questions for the reform of natural philosophy was very much on his mind. This is evident in his meditation on heresy. The essay expounds the words of Christ to the Sadducees, the arch-heretics of the Gospels, in Matthew 22:29: "Ye err, not knowing the scriptures nor the power of God." Bacon wrote:

> This canon is the mother of all canons against heresies. The cause of error is twofold: ignorance of the will of God, and ignorance or superficial consideration of the power of

---

not cite him until the end. See *Works of Lancelot Andrewes*, vol. 1, pp. 249–64, and 264–83. See also Lossky, *Lancelot Andrewes*, pp. 74, and 96–97, especially Lossky's footnote on p. 97, which examines Andrewes' direct use of Irenaeus' theology.

[18]   Benjamin Milner, "Francis Bacon: The Theological Foundations of 'Valerius Terminus,'" *Journal of the History of Ideas*, 58/2 (1997), pp. 245–64. See pp. 247–8.

God. The will of God is more revealed through the Scriptures: *Search the Scriptures*; his power more through his creatures: *Behold and consider the creatures*. So is the plenitude of God's power to be asserted, as not to involve any imputation upon his will. So is the goodness of his will to be asserted, as not to imply any derogation of his power.[19]

Throughout his philosophical writings Bacon relies heavily on the distinction he draws from Matthew 22:29 between the revelation of the will of God in the Scriptures and the revelation of His power in the creatures. As we will see later on, this distinction is often mistaken for a distinction between faith and science in Bacon's works. For now, we should note that heresy, according to the scheme which Bacon has set up, can potentially come from the misunderstanding of nature – God's creatures – as well as the more traditional explanation that it is the result of misunderstanding or misreading the Scriptures. The balance in wording of the last two sentences in this quotation is significant, for it sets the course for the rest of the essay on heresy. Reflecting his knowledge of the original meaning of heresy as choice of improper emphasis on one part of the faith, Bacon is seeking a balance which avoids emphasizing God's power over the goodness of His will, or emphasizing His good will at the expense of His absolute power. True religion must maintain this balance, or as Bacon says earlier in the essay, true religion is situated in the middle: "*Itaque religio vera sita est in mediocritate.*"[20]

After establishing the double source of heresy, Bacon left aside those heresies which resulted from confusion over God's will, and focused his attention on those which arise from denying God's absolute power. The most grievous denial of God's power is atheism, according to Bacon, after which come three lesser degrees of denial: the problem of dualism, in which an equal and opposite principle is opposed to God's goodness, followed by the neo-Platonic error of setting a privative principle, or a tendency toward dissolution in opposition to God's sustaining power, and, finally, the error of those who deny God's power by asserting that sin, at least, is *solely* the result of man's choice. Although the last idea would later be associated with Arminianism, in 1597 the works of Arminius were not yet available in England.[21] But, prior to Arminius, Calvin's doctrine had no shortage of objectors to whom Calvin had responded. It is clear from the text that Bacon's discussion is dependent on Calvin's *Institutes* and that, to this point, he is objecting to those whom Calvin

---

[19] For the *Meditationes Sacrae* the original Latin should always be consulted. This essay is found in WFB, vol. VII, pp. 240–42. The English translation is from the 1598 republishing of this volume in which the *Meditationes Sacrae* appeared in English (WFB, vol. VII, p. 252). The translator is unknown, and it is uncertain whether the translation had Bacon's approval. (Consider WFB, vol. VII, p. 229.)

[20] WFB, vol. VII, p. 241. In this light, Bacon's family motto, *mediocria firma*, represents not necessarily compromise, but the avoidance of error. This should certainly inform our reading of Bacon's advice to King James in *Considerations touching the better Pacification and Edification of the Church of England* where he advises James that as the "Christian moderator" he is "disposed to find out the golden mediocrity in the establishment of that which is sound, and in the reparation of that which is corrupt and decayed." He is not advising compromise, but rectitude (WFB, vol. X, p. 104.) Consider Deuteronomy 5:32.

[21] See Tyacke, *Anti-Calvinists*, pp. 4, 38–9, 65–6.

set up as targets in the *Institutes*. Bacon is not only borrowing from Calvin, he is responding to him, and his response makes it clear that he is distancing himself from Calvin no less than from Calvin's objectors:

> The third degree is of those who limit and restrain the former opinion to human actions only, which partake of sin: which actions they suppose to depend substantively and without any chain of causes upon the inward will and choice of man; and who give a wider range to the knowledge of God than to his power; or rather to that part of God's power (for knowledge itself is power) whereby he knows, than to that whereby he works and acts; suffering him to foreknow some things as an unconcerned looker on, which he does not predestine and preordain: a notion not unlike the figment which Epicurus introduced into the philosophy of Democritus, to get rid of fate and make room for fortune; namely the sidelong motion of the Atom; which has ever by the wiser sort been accounted a very empty device.[22]

Like Calvin, Bacon is concerned with safeguarding God's omnipotence. In a similar argument Calvin also made the point that God is not to be regarded as idle (*otiose*), but active in creation:

> And truly God claims, and would have us grant to him, omnipotence – not the empty, idle, and almost unconscious sort that the Sophists imagine, but a watchful, effective, active sort, engaged in ceaseless activity.[23]

Calvin is also interested in eliminating the philosophical concept of fortune or chance:

> ... there is no erratic power, or action, or motion in creatures, but that they are governed by God's secret plan in such a way that nothing happens except what is knowingly and willingly decreed by him.[24]

For both Calvin and Bacon, God is active in all parts of creation. For both Calvin and Bacon, there is nothing which happens which God has not ordained beforehand, and thus God's omnipotence is preserved. Beyond this, the similarities between the two end. They differ significantly with regard to the *way* in which God's omnipotence is exercised. For Bacon, God could "ordain" something beforehand without it being God's active decree. It could be approved because God has already woven it into a "bigger picture."

Bacon concluded his meditation on heresy as follows:

> But the fact is that whatever does not depend upon God as author and principle, by links and subordinate degrees, the same will be instead of God, and a new principle and kind of usurping God. And therefore that opinion is rightly rejected as treason against the majesty and power of God. And yet for all that it is very truly said that *God is not the author of evil*; not because he is not author, – but because not of evil.[25]

---

[22]   WFB, vol. VII, pp. 253–4.

[23]   ICR, Book I, ch. 16, sec. 3; Battles translation, p. 200.

[24]   Ibid., p. 201. Although this is spoken with reference to the heavenly bodies, Calvin employs it as a general principle which should ward off the superstition of astrology.

[25]   WFB, vol. VII, pp. 253–4.

For Bacon, God was author "by links and subordinate degrees," [*per nexus et gradus subordinatos*] which fits with his earlier caution against regarding sin as being "substantively and without any chain of causes" dependent on "the inward will and choice of man." What Bacon is arguing against here is a fairly extreme position, which would make sin fall outside of those things which are controlled and circumscribed by God's power. For Bacon, it is enough to recognize God as omnipotent if He governs the "chain of causes." God's actions do not have to be immediate in all things for Him to remain omnipotent. This runs counter to Calvin, who allowed for no action to occur in the world which was not directly and immediately God's action, whether in inanimate objects or in humanity. In regard to inanimate objects Calvin argued specifically against any chain of causes:

> And concerning inanimate objects we ought to hold that, although each one has by nature been endowed with its own property, yet it does not exercise its own power except in so far as it is directed by God's ever-present hand. These are, thus, nothing but instruments to which God continually imparts as much effectiveness as he wills, and according to his own purpose bends and turns them to either one action or another.[26]

In regard to creation, Calvin allowed no room for a distinction between God's "general providence," as the natural order of activity in the world, and His "special providence," by which He would circumvent the normal chain of causes and act more immediately. *All* of God's action was, for Calvin, a "special providence."[27]

When Calvin's understanding that God always acts immediately is applied to human actions, the result is that there is no room for any genuine free will. For Calvin, nothing occurs that God has not expressly and directly willed to occur. God does not merely *permit* sin, Calvin argued at length, but He Himself actively arranged all situations so that when sin occurs there is no alternative.[28] Although human beings are held responsible for willing to do evil, Calvin makes it clear that human will is so bounded by divine will that they were not able to do otherwise. Thus he argues that God does not merely desert the reprobate, turning them over to the temptation of Satan, but He establishes the condition in their hearts, which is responsible for their evil:

> But one can desire nothing clearer than where he so often declares that he blinds men's minds, smites them with dizziness, makes them drunk with the spirit of drowsiness, casts madness upon them, hardens their hearts. These instances may refer, also, to divine permission, as if by forsaking the wicked he allowed them to be blinded by Satan. But since the Spirit clearly expresses the fact that blindness and insanity are inflicted by God's just judgment, such a solution is too absurd. It is said that he hardened Pharaoh's heart, also that he made it heavy and stiffened it.[29]

---

[26]   ICR, Book I, ch. 16, sec. 2. Battles translation, p. 199.

[27]   ICR, Book I, ch. 16, sec. 4. Battles translation, pp. 201–3. While seeming, in the latter part of this *locus* to concede a universal providence, he accepts it only in so far as his objectors will concede to him that it only a matter of appearance: "*as if* [lat. *acsi*] they obeyed God's eternal command and what God has once determined flows on by itself" (p. 203).

[28]   ICR, Book I, ch. 18, sec. 1–3.

[29]   ICR, Book I, ch. 18, sec. 2. Battles translation, p. 231.

This argument, which presents human will as completely circumscribed and bound by the immediate actions of God pertains, in Calvin, to humanity after the fall.[30] Before the fall, Calvin claims, humans had "free will," but he does not discuss how this can be. For example, in the original sin, Adam and Eve could choose to eat of the forbidden tree according to their "free will," but at the same time they could not have chosen not to. God did not give them the gift of "perseverance" by which they could have avoided sin.[31] Calvin concludes that it is better not to inquire further into the matter.

Ultimately, God is for Calvin what Bacon cannot allow him to be: necessarily the author of evil. According to Book I, Chapter 18 of the *Institutes*:

> And now I have already shown plainly enough that God is called the Author of all the things that these faultfinders would have happen only by his indolent permission. He declares that he creates light and darkness, that he forms good and bad (Isa. 45:7); that nothing evil happens that he himself has not done (Amos 3:6).[32]

In Bacon's final word on the subject, that God is not the author of evil, he distances himself from Calvin's opinion. Bacon cannot conceive of God as the author of evil without "imputation upon his will." This would be a denial of the necessity of balancing God's power and His goodness, which he demanded at the beginning of his essay on heresies. As it became clear during the Arminian Controversy, there were many Calvinists who were far from comfortable with the radical determinism of the more literal adherents of Calvin's system, although they still referred to Calvin's writings to sort out the dilemma.[33] Bacon introduced the idea of a chain or order of causes, for which Calvin's system did not have room, to prevent God from being the author of evil. In so doing, Bacon was following the well-known argument of a writer who was also concerned with preserving both the goodness and the omnipotence of God – St Augustine.

In the fifth book of the *City of God*, Augustine argued against Cicero's understanding that a divine foreknowledge of the order of causes was antithetical to human free will. Augustine responded:

> But it does not follow that, though there is for God a certain order of all causes, there must therefore be nothing depending on the free exercise of our own wills, for our wills themselves are included in that order of causes which is certain to God, and is embraced by His foreknowledge, for human wills are also causes of human actions; and He who

---

[30]   See ICR, Book II, ch. 2.

[31]   ICR, Book I, ch. 15, sec. 8.

[32]   ICR, Book I, ch. 18, sec. 3. Battles translation, p. 233.

[33]   Thus "sublapsarianism" and "infralapsarianism" developed in opposition to the strict "supralapsarianism" of early Genevan Calvinism. Each group had a different understanding of the chronological order of God's decrees. For a concise discussion of the differences separating these groups see Philip Schaff, *History of the Christian Church* (New York, 1910) vol. 3, chapter 14. However, Thomas Cartwright, who was one of the champions of English Puritanism, took the more radical view. Consider Cartwright, *A Treatise of Christian Religion* (London, 1616), pp. 38-41. Bacon had read Cartwright at Gray's Inn, and probably met him at Cambridge. See Lisa Jardine and Alan Stewart, *Hostage to Fortune: The Troubled Life of Francis Bacon* (New York, 1998), p. 79.

foreknew all the causes of things would certainly among those causes not have been ignorant of our wills.[34]

"Foreknowledge" is not to be regarded as something apart from God's "supreme power" in this section or anywhere in the *City of God*, even as Bacon insisted that God's foreknowledge was not to be given a "wider range" than His power. After discussing the various types of causes in the order of causes, Augustine clarified this point:

> But all of them are most of all subject to the will of God, to whom all wills also are subject, since they have no power except what He has bestowed upon them. The cause of things, therefore, which makes but is not made, is God; but all other causes both make and are made.[35]

For Augustine, while humans could act and sin according to genuine free will, it was not possible for them to do anything contrary to, or even apart from, the *power* of God (not, notably, that humans could do nothing which was not the very *will* of God, as per Calvin). Sin did not develop in some vacuum of divine activity. Throughout the *City of God* the foreknowledge of God is connected to His power in such a way that nothing occurs which God has not foreknown, handled "in advance," and made part of his prearranged plan, according to his power:

> But because God foresaw all things, and was therefore not ignorant that man also would fall, we ought to consider this holy city in connection with what God foresaw and ordained, and not according to our own ideas, which do not embrace God's ordination. For man, by his sin, could not disturb the divine counsel, nor compel God to change what He had decreed; for God's foreknowledge had anticipated both, – that is to say, both how evil the man whom He had created good should become, and what good He Himself should even thus derive from him.[36]

Thus Augustine had no difficulty in making "sin" part of the greater plan of salvation, and part of the package which God "predestined."[37] But God is not responsible for the free choices of humans. He simply determined in advance that humanity should be allowed to abuse free will in light of the greater good which would come of it in

---

[34] Augustine, *De Civitate Dei*, V, 9. The Latin original may be consulted in *Corpus Christianorum*, vols 47–8 (Turnholtii, 1955). The translation here is that of Marcus Dods, in NPNF, series 1, vol. 2, p. 91.

[35] NPNF, p. 92.

[36] Augustine, *De Civitate Dei*, XIV, 11. Translation: NPNF, series 1, vol. 2, p. 271. See also *De Civitate Dei*, XII, 23.

[37] It is in light of the coming of Christ in time, which, according to Augustine, would not have occurred were it not for the Fall, that Augustine asks the rhetorical question: "...*nisi quia in eius aeternitate atque in ipso Verbo eius eidem coaeternato iam predestinatione fixum erat, quod suo tempore futurum erat?*" (*De Civitate Dei*, XII, 17, vol. 48, p. 373. Note: this occurs as section 16 of Book XII in the NPNF series. Augustine specifically describes that which occurs in time as "predestined" and "fixed" in the context of the entire process by which the Word, and eternal life, would be made real in time. Thus Bacon's usage of *praedestinet* (lit. to prearrange, or fix beforehand) and *preordinet* is in keeping with Augustine's language, while Calvin only applied "predestination" to the action of election.

the incarnation.[38] In the course of Augustine's discussion, the order of causes which includes genuine human choice serves a very specific purpose: it retains God's position as the omnipotent cause of all things, while moving the cause of evil down on the scale into the realm of human free will. Thus Augustine claimed:

> For, as He is the creator of all natures, so also is He the bestower of all powers, not of all wills; for wicked wills are not from Him, being contrary to nature, which is from Him.[39]

According to Augustine's order of causes, God can only be held accountable for sin in that he created people capable of sin, and he knew that it would inevitably happen. Thus God can be said to be the *ultimate* cause of sin, because He willingly created the situation in which He knew sin would happen. Although that makes God part of the chain, He is removed from being the *voluntary* cause of sin.[40] Evil (in the human sphere, not among angels) had as its source human choice:

> For God, the author of natures, not of vices, created man upright; but man, being of his own will corrupted, and justly condemned, begot corrupted and condemned children. ... And thus, from the bad use of free will, there originated the whole train of evils ...[41]

Augustine's use of the order of causes preserved the entire balance demanded by Bacon's discussion of heresies: God's absolute power was preserved, for He is never an idle observer. God's goodness was preserved, for He is not the voluntary cause of evil. It is also important to note that, according to Augustine, humanity did not suffer a sudden and complete loss of free will in the fall. In speaking of human free will Augustine always described something in continual existence.[42] This is a significant point of separation between Augustine and any form of Calvinism, for Calvin stated unequivocally that whatever free will man may be said to have had before the fall was taken from him afterward.[43]

---

[38] This concept is stated even more explicitly in *De Civitate Dei*, XXII, 1, where Augustine says that in spite of His knowledge that humans would use their free will to sin, "*nec illi ademit liberi arbitrii potestatem, simul praeuidens, quid boni de malo eius esse ipsi facturus.*" God, therefore, had already seen that human sin would be a part of His plan and He would make good from evil (*Corpus Christianorum*, vol. 48, p. 807).

[39] Augustine, *De Civitate Dei*, V, 9. Translation: NPNF, series 1, vol. 2, p. 92.

[40] Ibid.

[41] Augustine, *De Civitate Dei*, XIII, 14. Translation: NPNF, series 1, vol. 2, p. 251.

[42] Calvin's reading of Augustine, in ICR, Book II, ch. 2, is an attempt to make Augustine a supporter of Calvin's own extreme position, but it is not reflective of the consistency with which Augustine defends the concept of free will. It is, rather, an example of how Augustine could be turned toward many ends, as Gottschalk and Erigena had done some six centuries earlier. Calvin avoided mention of the sections that involve a chain of causes.

[43] ICR Book I, ch. 15, sec. 8. We should note that the separations in Calvinism which admitted for varying interpretations of the significance of pre-lapsarian free will ("supralapsarians," "infralapsarians," "sublapsarians," etc.), were concerned only with man's *original* free will, and all were agreed, against the Arminians, that this free will was subsequently lost. See Schaff, *History of the Christian Church*, vol. 3, chapter 14.

We can have no doubt that Augustine himself did not see the ambiguities in his system, which would plague subsequent Western theologians. Augustine's true position on the question of free will versus predestination has been the subject of fierce debate for many centuries. Both sides of the ninth-century predestinarian controversy could appeal to Augustine for support,[44] and quotations from Augustine saturate the Calvin's writings as well. However, with later thinkers, it is always worth noting which passages they choose in Augustine to support their particular points. Bacon has, amidst Augustine's ambiguities, attached himself to an argument used for the specific purpose of carving out a space for genuine human free will, both before and after the fall.

Bacon's Instauration, which was very much on his mind while he was writing the *Meditationes Sacrae*, was to be a human project, requiring the power or industry of man, and genuine human agency. As Karl Wallace has observed, the distinctive feature of Bacon's understanding of the human will was the "power of choice."[45] Throughout the *Instauratio Magna* Bacon's new method is presented as an alternative to the old way of error, which would require a free choice on the part of his readers even as the old way had been the result of man actively placing his trust in erroneous methods. The key to the advancement of the science was making this choice. As Bacon claimed in *Valerius Terminus*, man could obtain comprehension of the entire created order, if he would act on his own divinely given power: "if man will open and dilate the powers of his understanding as he may."[46] Augustine, who made human free will an essential part of the order of causes, was compatible with other currents of Bacon's thought at this time, while Calvin, who denied any genuine human agency after the fall, was not.

For Augustine, God's omnipotence functioned in such a way that God was never separate from the motions of His creation: He was never idle but always actively involved, even when humans were making free choices. In a similar way, Bacon conceived his Instauration as both a work of God and a work of human agency and achievement, based on the differing places of God and man in the order of causation. In the *Novum Organum*, when Bacon is presenting the reasons why his readers should adopt a hopeful outlook for the success of his project, he gives both a divine and a human reason for this hope: first, God, in His providence, has already set the Instauration in motion, and God will always bring His own works to completion; second, and immediately following this point, the Instauration had not occurred in the past because mankind has not tried it before, but there was nothing in nature itself which prevented it, and now it will succeed if the past errors are corrected, as it is within the scope of human power to do.[47]

Most of the *Meditationes Sacrae* reflects ideas common to all forms of Christianity, although there is a clear preoccupation with the relationship between

---

[44]   See Jaroslav Pelikan, *The Christian Tradition*, (5 vols, Chicago, 1978), vol. 3, pp. 80–98.

[45]   Karl Wallace, *Francis Bacon on the Nature of Man* (Champaign–Urbana, 1967), p. 140.

[46]   WFB, vol. III, p. 221.

[47]   See Aphorisms XCIII and XCIV respectively in WFB, vol. I, p. 200. For translation see WFB, vol. IV, pp. 91–2.

God and nature.[48] However, there is further evidence in the *Meditationes* of a movement in Bacon's thought away from Calvinism and toward the perspectives of Christian antiquity which would become more pronounced in later writings. In an essay on hypocrites, Bacon writes the following concerning monasticism, which, by Protestant consensus, had come to embody the spiritual pomp and arrogance which were the hallmarks of hypocrisy:

> By which error the life monastic was, not indeed originated (for the beginning was good), but carried into excess. For it is rightly said *that the office of prayer is a great office in the Church;* and it is for the service of the Church that there should be companies of men relieved from cares of the world, who may pray to God without ceasing for the state of the Church. But this institution is a near neighbour to that form of hypocrisy which I speak of: nor is the institution itself meant to be condemned; but only those self-exalting spirits to be restrained.[49]

Calvin also claimed that monasticism was good in its original form, but for a very different reason. In Book Four of the *Institutes*, Calvin denounced the recent forms of monasticism because they had wandered from their original purpose, which was to provide the Church with trained and pious clergy.[50] Calvin acknowledged that the majority of monastics did not move on to "greater offices" [*ad maiora munera*], and never intended to, but he contended, nevertheless, that early monastic communities were defensible for precisely this purpose. In this way he could exonerate some of his favorite sources, Augustine and Chrysostom, as well as the Cappadocian fathers, for whom monasticism was an important and necessary institution. For Bacon, monasticism was a valuable institution in and of itself, for the same reason which the early monastics themselves gave, namely that there should be a class or order within the Church dedicated to seclusion and a life of perpetual prayer. The value of monasticism in the Church for Calvin was time-bound: its role had been supplanted by seminaries. By contrast, the value of monasticism in the Church for Bacon was intrinsic. Bacon's perspective on monasticism was not entirely foreign to Protestantism,[51] but it differs significantly from that of Calvin. Although monasteries were a thing of the past in Tudor England, there were still those like Bacon's close friend Lancelot Andrewes who lived and valued a life of pious and chaste seclusion.

---

[48]   For example, Bacon begins the *Meditationes* with a brief consideration of the works of God and the works of man, then discusses the significance of the miracles of Christ to the laws of nature which God had established in creation. This is followed by a meditation on the innocence of the dove and the wisdom of the serpent (from Matt. 10:16), in which Bacon defends the idea of experimental knowledge from its pious detractors (WFB, vol. VII, pp. 233–5).

[49]   WFB, vol. VII, p. 238; translation, p. 249.

[50]   ICR, Book IV, ch. 13, sec. 8.

[51]   Martin Luther's early tract, *de Votis Monasticis*, also allowed that the monastic life could be intrinsically good, if it were a matter of free choice, not regarded as an inherently superior state, and not made compulsory through vows. See Martin Luther, *Luther's Works*, ed. Jaroslav Pelikan (St Louis, 1955), vol. 44, pp. 221–400. Among Luther's main charges against monastic vows are the concerns which Bacon also expresses: that they lead to hypocrisy and spiritual arrogance (p. 280). Bacon put a much greater stress on the value of monasticism for the Church as a whole than did Luther, who saw it as unnecessary.

Bacon's *Confession of Faith*

The clearest evidence in Bacon's own texts of a profound movement away from Calvinism during the late 1580s and early 1590s is his *Confession of Faith*. The dating of the *Confession* is uncertain, but the earliest extant manuscript copy, written in the hand of a secretary, ascribes it to "Mr." Bacon. As Spedding noted, this would place it prior to his knighthood in 1603.[52] The language is polished and the thoughts are carefully organized in the manuscript edition, suggesting that Bacon had been working these ideas out for some time. The doctrinal content itself shows that he had moved beyond the fairly standard formulations of the *Meditationes Sacrae*, and had settled on a direction which no longer attempted to reconcile his thought with the concerns of Calvinism. It is possible that the *Meditationes* represents one more politically acceptable way in which Bacon was handling key theological issues of his day, and the *Confession* represents another more private line of thought. Given the manner in which the key issues of predestination and free will are treated, it seems more appropriate to regard this as the culmination of a progression away from Calvin via the middle stage of the *Meditationes Sacrae*. The *Confession* presents a theological system which is entirely irreconcilable with Calvinism on a number of distinctive doctrines, but none of its elements is without precedent in Christian history. Bacon was willing to look to a wide variety of Christian authorities to find answers which were more suitable than those of Calvin.

A Confession of Faith is set forth as Francis Bacon's personal creed. It begins with the words of the creedal formula "I believe," and it discusses the matters of the faith according to the standard pattern of the "three articles" of both the Apostles' and the Nicene Creeds. These articles address God's work in creation, the incarnation, and sanctification respectively. The threefold order of the Articles of Faith could easily be regarded as part of the standard mental furniture of all those raised in the Christian faith up through Bacon's time, solidified in Protestant circles by Luther's *Catechisms* and the *Heidelberg Catechism*. James Spedding used his introduction to Bacon's *Confession* as a platform to argue the sincerity of Bacon's Christianity, taking a stand against the association of Bacon with Enlightenment atheism. In the process he suggested that Bacon's personal "creed" may be understood in terms typical of Reformed theology: "but the entire scheme of Christian theology, – creation, temptation, fall, mediation, election, reprobation, – is constantly in his thoughts … "[53] Spedding is right about the dominance of these doctrines for Bacon, but there is a greater significance in this text than he realized. The doctrines in the *Confession* would have raised serious concerns among the Reformed theologians of Bacon's day had he gone so far as to publish them.

Immediately after asserting the eternality and goodness of the Trinity in the first article, Bacon presents a doctrine of Christ as "Mediator" which would have drawn heavy fire from most contemporary Protestants:

---

[52] WFB, vol. VII, p. 216.

[53] Ibid., pp. 215–16.

> I believe that God is so holy, pure, and jealous, as it is impossible for him to be pleased in any creature, though the work of his own hands; So that neither Angel, Man, nor World, could stand, or can stand, one moment in his eyes, without beholding the same in the face of a Mediator; And therefore that before him with whom all things were present, the Lamb of God was slain before all worlds; without which eternal counsel of his, it was impossible for him to have descended to any work of creation; but he should have enjoyed the blessed and individual society of three persons in Godhead only for ever.

> But, that out of his eternal and infinite goodness and love purposing to become a Creator, and to communicate with his creatures, he ordained in his eternal counsel that one person of the Godhead should in time be united to one nature and to one particular of his creatures: that so in the person of the Mediator the true ladder might be fixed, whereby God might descend to his creatures, and his creatures might ascend to God.[54]

According to Bacon, God "chose (according to his good pleasure) Man to be that creature, to whose nature the person of the eternal Son of God should be united."[55] Notably, the idea of Christ being the intermediary between God and creation is not tied as a matter of necessity to the fall. The purpose of the hypostatic union, the uniting of the human and divine natures in the person of Christ, is not first and foremost to rescue human beings from sin, but to unite God and His creation so that there may be communication between them. Bacon is espousing a doctrine of deification. Humanity's fall was known to God, and as such He made it a part of His plan to unite human and divine natures, but the fall itself was not an essential part of the plan. The fall was a twist contributed by humans, yet foreknown:

> That he made all things in their first estate good, and removed from himself the beginning of all evil and vanity into the liberty of the creature; but reserved in himself the beginning of all restitution to the liberty of his grace; using nevertheless and turning the falling and defection of the creature, (which to his prescience was eternally known) to make way to his eternal counsel touching a Mediator, and the work he purposed to accomplish in him.[56]

The Mediator is the means and the intermediary (truly the *media*) by which God interacts with creation, and the office was necessary for the very act of creation to occur:

> That by virtue of this his eternal counsel touching a Mediator, he descended at his own good pleasure, and according to the times and seasons to himself known, to become a Creator; and by his eternal Word created all things, and by his eternal Spirit doth comfort and preserve them.[57]

The union of God and creature in the Mediator, now that sin had entered the world, was also the means whereby salvation would come to the Church:

> ... he chose (according to his good pleasure) Man to be that creature, to whose nature the person of the eternal Son of God should be united; and amongst the generations of men,

---

[54]   Ibid., p. 219.
[55]   Ibid., p. 220.
[56]   Ibid.
[57]   Ibid.

elected a small flock, in whom (by participation of himself) he purposed to express the riches of his glory; all the ministration of angels, damnation of devils and the reprobate, and universal administration of all creatures, and dispensation of all times, having no other end, but as the ways and ambages of God to be further glorified in his Saints, who are one with the Mediator, who is one with God.[58]

Although some of God's creatures, having fallen from perfection, would not be restored, the saints who were in the Church were to be glorified by participation with God through their unity with the Mediator. The work of Christ on the cross was important, in that sin required a sacrifice as payment,[59] but this was only one aspect of the incarnation. In the end, the incarnation served as the means by which God would accomplish unity with his creation through his appointed Mediator. The theology which is expounded throughout these passages is certainly not without precedent in Christian history. Similar discussions of the Logos as an intermediary between God and creation can be found throughout the first four centuries of Christianity, but no such discussion can be found in Calvin's *Institutes*. What Bacon has expressed in these passages was explicitly condemned by Calvin in his attack on the Lutheran theologian, Andreas Osiander.

For Calvin, the only reason that God united with man in the person of Christ was to rescue man from sin. In the section of the *Institutes* on Christ as Mediator, Calvin wrote:

> Scripture universally assigns no other end, for the Son of God voluntarily assuming our flesh, and also accepting it as a mandate of the Father, except to become a victim to placate the Father to us.[60]

It was this act of propitiation for which Christ was properly called "mediator." Calvin devoted a great deal of space to rejecting the "vague speculations" (*vagas speculationes*) of others who might suggest that there was more to the doctrine of the incarnation than the work of redemption from sin. Calvin's main target was Andreas Osiander. In his publication of 1551, *Von dem Einigen Mitler Jhesu, Christo und Rechtfertigung des Glaubens,* Osiander had proposed a mystical view of the incarnation in which sin was incidental to the union of God and man in Christ. Justo Gonzalez has aptly summarized Osiander's understanding of Christ's mediatorial role:

> Adam is said to have been made after the image of God, because before the foundation of the world God had decided that the Son was to become incarnate. Thus the incarnation was not God's response to sin, but his eternal purpose. Even if Adam had not fallen, Christ would have become incarnate. But, even before the incarnation, humankind was created so that the image of God – that is, the Son – could dwell in it.[61]

---

[58] Ibid.

[59] Ibid., p. 223.

[60] ICR, Book II, ch. 12, sec. 4. Translation my own.

[61] Justo L. Gonzalez, *A History of Christian Thought* (3 vols, Nashville, 1975), vol. 3, p. 104. Gonzalez has an accessible presentation of Osiander's theology in the English language, but a more thorough discussion is that of Emanuel Hirsch, *Die Theologie des Andreas Osiander* (Gottingen, 1919).

For Calvin, Christ was Mediator because he was Redeemer. But Osiander added a dimension to the role of Christ as Mediator, which office applied primarily to the unification of God and humanity, and secondarily to the redemption of humans from their sinful state. As Gonzalez put it, "Because of the fall, the incarnation took on an additional purpose: the redemption and justification of humankind."[62] According to Osiander, Christians were justified before God, and hence saved from destruction, because Christ was mystically united to them and dwelling in them. When God beheld the individual Christian he beheld the person of the Son, united to an individual who was in the process of being perfected, or deified. Calvin repeatedly attacked Osiander for raising an old speculation: "that Christ would still have become man even if no means of redeeming mankind had been needed" [*Christum, etiam si ad redimendum humanum genus non fuisset opus remedio, futurum tamen fuisse hominem*].[63] But this hypothetical question is not what Calvin found most objectionable in Osiander; it was the theology behind it, which interpreted Christ's mediation as entailing a cosmic significance apart from propitiation for sin. On this point, Bacon's *Confession of Faith* stands equally condemned by the *Institutes*.

Osiander's mystic/neo-Platonic interpretation of the incarnation reflected one trend in the theology of the early Reformation on the continent. This trend was entirely incompatible with Reformed theology as expressed by Calvin, and it also gradually fell out of favor among the Lutherans as well. By 1577 Osiander had been condemned by name in the Lutheran confessional document, the *Formula of Concord* because his understanding of salvation as deification was regarded at this time as being at odds with the key Reformation doctrine of "forensic justification" – the idea that humanity, while still sinful, is "declared righteous" by a decree of God in light of the propitiatory work of Christ's death.[64] To Lutherans as well as Calvinists in the later sixteenth century, Osiander and those who thought like him came to be regarded as having views that were far too close to the Roman Catholic understanding of grace being *imparted* to man (through the indwelling of the Son) rather than *imputed*, as the doctrine of forensic justification maintained.

Bacon did not specifically raise the hypothetical question of what would have occurred had man not sinned, but, in other respects, the similarities between Bacon's system and Osiander's are manifest. Given the limited influence of Lutheranism in England, as well as the negative response to Osiander's theology by Bacon's day, we cannot be sure Bacon ever read his work. The Bodleian Catalogue of 1605 does not list any of Osiander's controversial works as being in the library at Oxford.[65] But, as Calvin noted, Osiander was not unique. Osiander and Bacon both had access to the same Eastern Church Fathers who held that deification was the central doctrine

---

[62]   Gonzales, *History of Christian Thought*, p. 104.

[63]   ICR, Book II, ch. 12, sec. 4, p. 467.

[64]   Karl Barth's discussion of the incompatibility of forensic justification and "an essential deification of man" is an accurate summary of late sixteenth-century opinion on the matter. See Karl Barth, *Church Dogmatics*, trans., G.T. Thomson (Edinburgh, 1936), pp. 274–5.

[65]   The only work in the Bodleian in 1605 was Osiander's Greek and Latin harmony of the Gospels. Cf. Thomas James, *Catalogus Librorum Bibliothecae Publicae quam vir Ornatissimus Thomas Bodleius ...* , (London, 1605), p. 106.

of the Christian faith. Bacon's use of Revelation 13:8 – "the Lamb slain before the foundation of the world" – to establish the centrality of the incarnation in God's eternal plan is strikingly similar to the use of the same verse by Maximus the Confessor (d. 662),[66] but it is doubtful whether Bacon had direct access to Maximus'[67] whose ideas were available through the writings of John Scotus Erigena, Maximus' medieval translator. Erigena used Maximus as the starting point for much of his own doctrine, which he took in some famously heterodox directions.[68] Furthermore, Erigena's own reputation would probably have colored Maximus very negatively. By contrast, Irenaeus of Lyons was a far more ancient, and hence purer, source, whose essential orthodoxy was acknowledged by all. The similarities between Bacon's *Confession* and the fourth and fifth books of Irenaeus' *Adversus Haereses* are striking, and Irenaeus was readily available in Bacon's England.

In 1526 the first printed edition of Irenaeus' five books against the Gnostic heresies came forth from the editorial hand of Desiderius Erasmus. This initiated a wave of scholarly editions of Irenaeus, which would swell over the course of the sixteenth century.[69] It is easy to understand Irenaeus' popularity during the Reformation. Writing in the second century, Irenaeus was impressively close to the apostolic age of Christianity, and he had left a detailed exposition of the faith from that time. The thoroughness of Irenaeus' discussion is due to the subject matter of the five books – a defense of orthodox Christianity against the various heresies of Christian Gnosticism. As a voice from a more ancient and more Eastern form of Christianity, Irenaeus laid down a challenge to all sides.[70] In his writings Protestants found ideas to counter contemporary Roman Catholicism, while Catholics could point to Irenaeus' emphasis on tradition and Apostolic Succession as evidence that the Protestants had rejected certain central ideas of early Christianity. In *Adversus Haereses* early modern Christians were confronted with a manner of expressing the faith and a worldview which had grown foreign with the passage of time. Now it had to be respected and ,as many felt, assimilated, by virtue of the authority of the source.

---

[66] Cf. Vladimir Lossky, *The Mystical Theology of the Eastern Church* (Crestwood, NY, 1976), pp. 137–8.

[67] Lancelot Andrewes was well-read in the available Greek sources and tended to mention them when using their ideas, but never mentions or cites Maximus. The Bodleian Catalogue lists the works of a St Maximus, but it is likely that this was Maximus of Turin, who was known at this time and cited by Andrewes, among others.

[68] Erigena concluded, with Origen, that there was neither hell nor the punishment of the wicked. See Pelikan, *The Christian Tradition*, vol. 3, p. 104. He also argued, from the immanence of God, that there was no special presence of Christ in the Sacrament of Holy Communion (ibid., p. 96.)

[69] The textual history of Irenaeus in the early modern period is to be found in Migne's *Patrologia Graeca*, (Paris, 1892), vol. 7, col. 1–23.

[70] Although Irenaeus served as bishop in Gaul, his worldview and theology were Hellenic. He came from Asia Minor, wrote, originally, in Greek, and gives evidence throughout his writing of the influence of Hellenic philosophy. Southern Gaul itself, and Lyons in particular, was culturally impacted by migrations of Greeks, and was a mission field of the Christians of Asia Minor. See Robert M. Grant, *Irenaeus of Lyons* (London, 1997), pp. 4–5.

The doctrine of the Word of God, the Logos, as mediator between God and creation is a central aspect of Irenaeus' Christology. As in Bacon's *Confession*, the central point of the incarnation is not redemption from sin, but the perfection of the communication between God and man. According to Irenaeus:

> Now this (that by which God created all things) is His Word, our Lord Jesus Christ, who in the last times was made a man among men, that He might join the end to the beginning, that is, man to God. Wherefore the prophets, receiving the prophetic gift from the same Word, announced His advent according to the flesh, by which the blending and communion of God and man took place according to the good pleasure of the Father, the Word of God foretelling from the beginning that God should be seen by men, and hold converse with them upon earth, should confer with them, and should be present with His own creation, saving it, and becoming capable of being perceived by it, and freeing us from the hands of all that hate us, that is, from every spirit of wickedness; and causing us to serve Him in holiness and righteousness all our days, in order that man, having embraced the Spirit of God, might pass into the glory of the Father.[71]

The ultimate goal for Irenaeus is the mystical union of God and man. As a result of the fall, salvation from sin was also an important aspect of the incarnation. But the fall itself was an act of human free will which God, by virtue of his foreknowledge, worked into his equation:

> For after His great kindness He graciously conferred good (upon us), and made men like to Himself, (that is) in their own power; while at the same time by His prescience He knew the infirmity of human beings, and the consequences which would flow from it; but through (His) love and (His) power, He shall overcome the substance of created nature.[72]

Irenaeus, like Bacon, never raised Osiander's *contra facta* hypothesis of the necessity of the incarnation had humans not sinned. But the salvific aspect of the incarnation is always subsumed under the larger divine plan to unite God and creation. For Irenaeus, humankind was the specific creature to which the second person of the Trinity should be united in time, and hence had special status. But, like Bacon again, Irenaeus used humankind as the point of contact by which God would, through the mediating actions of the incarnate Christ, be in communication with the entire cosmos:

> For the Creator of the world is truly the Word of God: and this is our Lord, who in the last times was made man, existing in this world, and who in an invisible manner contains all things created, and is inherent in the entire creation, since the Word of God governs and arranges all things; and therefore He came to His own in a visible manner, and was made flesh, and hung upon the tree, that He might sum up all things in Himself.[73]

---

[71]   *Adversus Haereses*, Book 4, ch. XX, 4. For the critical Latin edition see the *Sources Chrétiennes*. The Greek of certain passages has been preserved only in secondhand quotations of certain later Fathers. The English translation is taken from the *Ante-Nicene Fathers* (ANF), vol. 1. The numeration of chapters and paragraphs in the *Ante-Nicene Fathers* is used here, along with page numbers from ANF, but the numeration has not been standardized and varies from version to version.

[72]   *Adversus Haereses*, Book 4, ch. XXXVIII, 4 (ANF vol. 1, p. 522).

[73]   Ibid., Book 5, ch. XVIII, 3 (ANF, vol. 1, pp. 546–7).

According to Irenaeus, the incarnation was the ultimate fulfillment of the role played by the Logos from the very beginning as the mediator, or "go-between," bridging God and creation. The incarnation had to occur for the connection between God and creation to be complete. The creatures, without the benefit of Christ coming in the flesh to unite God and creation, could neither comprehend nor communicate with God:

> For in no other way could we have learned the things of God, unless our Master, existing as the Word, had become man. For no other being had the power of revealing to us the things of the Father, except His own proper Word.[74]

The Logos is always the intermediary between God and creation – it was *by* the Word that all things were created, and it was the second person of the Trinity who communicated with Moses and Abraham.[75] At the very center of this mediating activity, and as the purpose for creation itself, is the event of the incarnation, by which the true communication between God and humanity is established. Without the incarnation, God is incomprehensible, and his power and glory are holy and unapproachable, for "no man shall see God and live." But in the incarnation God presents Himself in a form accessible to humans, and through this incarnate form they gain immortal life and the ability to "pass into the glory of the Father."[76] Similarly, the love of the Father for humankind is the result of the incarnate Logos as a mediator:

> And then, again, this Word was manifested when the Word of God was made man, assimilating Himself to man, and man to Himself, so that by means of his resemblance to the Son, man might become precious to the Father.[77]

This passage provides an authoritative precedent for Bacon's statement that "God is so holy, pure and jealous" that nothing "could stand, or can stand one moment in his eyes, without beholding the same in the face of a Mediator."

The doctrine of the Mediator is only one of many elements of Bacon's *Confession of Faith* which run contrary to Calvinist theology. In the *Confession* Bacon went much further than in the *Meditationes Sacrae* in his rejection of Calvin's determinism. First, he explicitly stated that the fall was entirely the result of human free will: God "removed from himself the beginning of all evil and vanity into the liberty of the creature."[78] We should note here that Bacon was very careful to avoid the time-bound language which forced Calvin to his conclusions in the first place. There is no more discussion of "*pre*destination" and "*fore*ordination," as in the *Meditationes Sacrae*. Instead, Bacon referred to God's "eternal" will. Humankind is bound to linear time, and to perceiving things according to linear time. God is eternal, and transcends time. It is still necessary from a human, time-bound perspective, to use

---

[74]   Ibid., Book 5, ch. I, 1 (ANF, vol. 1, p. 526).

[75]   Ibid., Book 4, ch. XII–XIII (ANF vol. 1, pp. 475–8); also ch. XX, 9 and 10 (ANF, vol. 1, pp. 490–1).

[76]   Cf. ibid., Book 4, ch. XX, 4 (ANF, vol. 1, p. 488).

[77]   Ibid., Book 5, ch. XVI, 2 (ANF, vol. 1, p. 544).

[78]   WFB, vol. VII, p. 220.

words like "prescience," as Bacon did to explain that the fall was not a surprise to God, but he qualifies it immediately as what is "*eternally* known:"

> ... using nevertheless and turning the falling and defection of the creature, (which to his prescience was eternally known) to make way to his eternal counsel touching a Mediator ...[79]

By refusing to bind the knowledge or will of God to the temporal categories of "before" and "after," Bacon, unlike Calvin, again followed the practice of the early Church. Consequently, he could preserve the idea of a truly free will in both God and humankind without calling God's power into question. For Bacon, as for Irenaeus, Augustine, and many other Church Fathers, human free actions were not only eternally known, but also eternally accounted for by God. The relationship between God and man is conceived much as if it were a cosmic dance between two freely willing partners. One of the partners, however, transcends time, whereas the other is bound to it. God not only already knows the missteps of human beings, but carries on the dance in such a way that these are woven in, and turned toward a good end.

Quite naturally, this approach has significant ramifications for the doctrine of "election," according to which God has chosen, in Bacon's words, "a small flock" for salvation. Bacon avoids the Calvinist move from the idea of "election" to the "predestination" of certain *individuals* to heaven or hell as if human free will were not operative. Yet, later in the *Confession*, Bacon also acknowledges that the names of those who are to be saved are "already written in the book of life."[80] In the complex interaction between the time-bound and the transcendent, such paradoxes abound as a result of the time-bound nature of human language, and should not be taken as an indication of a specific doctrine of predestination. Augustine wrestled with the limitations of language in regard to this very issue.[81] The resolutions can be dizzying, and, apart from the internecine fighting of the Reformed in the Arminian controversy, logical resolutions were widely regarded as pointless in the early modern period, because they deal with matters beyond the comprehension of the human mind. Resolution would require the time-bound creature to be capable of perceiving salvation history from the perspective of timelessness. Thus, for Bacon, the names of those who chose properly in life and joined the small flock are in the Book of Life because that book transcends time. Philip Melanchthon, a key theologian who explicitly rejected the idea of double predestination, established a common pattern of this explanation in his *Loci Communes,* of 1543. Those who are in the chosen flock who "hear the voice" of the Shepherd are elect because God "gives approval to and elects those who are obedient to His call."[82]

There are still other elements of the *Confession of Faith* that are incompatible with Calvin's distinctive doctrines. As in the *Meditationes*, Bacon again denied Calvin's doctrine that God's governance of creation was immediate, rather than through a

---

[79]   Ibid.

[80]   Ibid., p. 225.

[81]   See Augustine's qualification of his own usage in *De Civitate Dei*, XII, 25: *Sed ante dico aeternitate, non tempore. Quis enim alius creator est temporum, nisi qui fecit ea, quorum motibus currerent tempora?*

[82]   Philip Melanchthon, *Loci Communes*, trans. J.A.O. Preus (St Louis, 1992). p. 174.

chain of causes: "yet nevertheless he doth accomplish and fulfill his divine will in all things great and small ... though his working be not immediate and direct, but by compass; not violating Nature, which is his own law upon the creature."[83]

Another example of divergence from Calvin is Bacon's claim that, in the incarnation, Christ "accomplished the whole work of the redemption and restitution of man to a state superior to the Angels, whereas the state of his creation was inferior; and reconciled or established all things according to the eternal will of the Father."[84] Calvin explicitly denied that the redeemed state would be in any way superior to that of the angels and, notably, he made this statement as a conclusion to a section directed against Osiander:

> But it cannot be denied that the angels also were created in the likeness of God, since, as Christ declares (Mt. 22: 30), our highest perfection will consist in being like them.[85]

Again, the concept that human beings, in their redeemed state, are superior to the angels is a commonplace for Bacon, Andrewes, Osiander, and Irenaeus. In the very last line of the last book in *Adversus Haereses* Irenaeus left his readers with the following thought:

> For there is the one Son, who accomplished His Father's will; and one human race also in which the mysteries of God are wrought, "which the angels desire to look into;" (1 Pet. 1: 12) and they are not able to search out the wisdom of God, by means of which His handiwork, confirmed and incorporated with His Son, is brought to perfection; that His offspring, the First-begotten Word, should descend to the creature, that is, to what had been moulded, and that it should be contained by Him; and, on the other hand, the creature should contain the Word, and ascend to Him, passing beyond the angels, and be made after the image and likeness of God.[86]

Irenaeus is one of the earliest and strongest sources for an explicit and systematic doctrine of deification in which the incarnation served the ultimate divine end of unifying God and creation. Bacon's own doctrine of deification follows that of Irenaeus in all particulars.

There are several other doctrines in Bacon's *Confession* which make it clear that Bacon, like Andrewes, aligned himself with the Christianity of late antiquity, and not the Reformation of Luther or Calvin. Even as Irenaeus regarded the Apostolic Succession as an indispensable mark of the Church, Bacon put a special emphasis on the "holy succession" of clergy which united the Church "from the time of the apostles and disciples which saw our Saviour in the flesh unto the consummation of the work of the ministry."[87] While Calvin never objected to the concept of the Apostolic Succession as it was understood by the early Church, he also did nothing with it himself, other than to caution against it because of the use which the papacy

---

[83] WFB, vol. VII, p. 221.

[84] Ibid., p. 223.

[85] ICR, Book I, ch. 15, sec. 3 (trans. Beveridge).

[86] *Adversus Haereses*, Book 5, ch. XXXVI, 3 (ANF, vol 1, p. 567).

[87] WFB, vol. VII, p. 225. Cf. *Adversus Haereses*, Book 3, ch. 3 (ANF, vol. 1, pp. 415–16).

made of it.[88] Likewise, Bacon carefully defended the doctrine that "the blessed Virgin may be truly and catholicly called *Deipara*, the Mother of God," against the suggestion that she was merely the mother of the human nature in Christ.[89] By the time of the Reformation no mainline group, Protestant or Catholic, denied this, and it was not one of the hot issues of the day. In the early fifth century, however, it was the occasion for tremendous debate, which ended with the condemnation of Nestorius (who denied that Mary was the Mother of God) at the Third Ecumenical Council held at Ephesus in 431. No Protestant would have been bothered by Bacon's discussion of this point other than, perhaps, to wonder why Bacon thought that this was still an issue requiring special mention. In this light, we may recognize this as further evidence that Bacon, in his *Confession*, was concerned with setting forth the faith as it was formulated in the first centuries of Christianity.

The period when Bacon was most diligently working out his plan for the reform of learning and natural philosophy was also the period when he departed completely from Calvinism for the theology of the Church Fathers, particularly that of Irenaeus of Lyons. These two phenomena must be considered together. Bacon's turn toward the ancient formulations of Augustine and Irenaeus is associated with the question of how the Creator relates to creatures, as well as the question of human potential. In the ancient fathers Bacon found answers which would transform his understanding of sacred history serve as the theological undergirding for the Great Instauration. And Bacon was not alone as he wrestled with the Patristic understanding of God and nature.

Bacon's theology had come to resemble that of his friend, Lancelot Andrewes, and the exchange of ideas was by no means one-way. During the early years of his friendship with Bacon, Andrewes was busy lecturing on the book of Genesis, and on the creation and fall narratives of the first chapters in particular. His lectures were well attended, and his students took copious notes, which, some sixty-odd years later were edited and compiled in a thick volume entitled Αποσπασματια *Sacra*. Significantly, this book is filled with discussions of natural philosophy. It is also striking that the points made by Andrewes in these lectures neatly match up with the points made by Bacon regarding the creation and the fall. The friendship between Bacon and Andrewes in these early years appears to have been marked by an ongoing discussion of themes which would become prominent in Bacon's later writings on the Great Instauration.

---

[88]   ICR, Book IV, ch. 2.
[89]   WFB, vol. VII, 223.

Chapter 3

In the Beginning: The Creation of Nature and the Nature of the Fall

By the time King James came to the throne in 1603, Francis Bacon had spent much of his adult life trying to gain a position at court. More to the point, as far as Francis was concerned, he was trying to gain the type of position or patronage which would allow him to devote himself to his program for the reform of natural philosophy, as the 1592 letter to his uncle, Lord Burghley, indicates. With the accession of James, Bacon met with a breakthrough on this front. He was officially named the King's "Learned Counsel," and was knighted in July of 1603. From this point onward, Bacon would ascend through the positions of Solicitor-General (1607) and Attorney-General (1612) to the precarious height of Lord High Chancellor of England (1618), at which point he was made Baron Verulam, and Viscount St Albans the next year. Bacon's political ambition may have led him to overshoot the goal which he had stated to Lord Burghley. But it may also be that, in pursuing ever higher offices, Bacon desired the type of position in which his plans for a new method in natural philosophy could be put into practice more effectively and more officially. For Francis Bacon the reform of learning was not a secular pursuit, but a divine mandate.

During his rise to power in the early years of the seventeenth century, Bacon finally began publishing his program for the reform of learning, which he would place under the heading of *Instauratio Magna*, or "Great Instauration." The term "Instauration" did not refer specifically to Bacon's program for reform, or to the empirical method contained in his writings. The "Instauration" was an event which had a unique place in the narrative of sacred history.

The Instauration as an Event in Sacred History

Charles Whitney has examined the meaning of the term, *Instauratio*, which Bacon applied to the event.[1] The word itself is theologically charged. Although it could be translated as "restore," "re-establish," "renew," or "begin again," and it could refer to many acts of renovation, it was also a word characteristically associated with the re-establishment of religious rites in the classical world. There are also architectural overtones to the word, but even these are theologically conditioned. In the standard Vulgate translation of the Old Testament, *instauratio* referred specifically to the rebuilding of the temple on the return from the Babylonian captivity. Then, through

---

[1]   Charles Whitney, "Francis Bacon's Instauratio: Dominion of, and over, Humanity," *Journal of the History of Ideas*, 50/ 3 (July–September 1989), pp. 371–90.

the tremendous influence of Augustine, *instauratio* came to signify "the new covenant" specifically, and when it was applied to the individual, "the instauration of the new man is signified by the resurrection."[2] Whitney also observed that, in one of the central verses on which Irenaeus based his doctrine of recapitulation, Ephesians 1:10, the word is used in the Vulgate as a translation for "summed up." Instead of Christ "summing up" all things in himself, as we observed Irenaeus saying in the last chapter, the Vulgate reads that all things are "instaured" in Christ.[3]

The event of the Instauration, according to the word itself, was a divine action of restoration which could not be dissociated from its implications in theological Latin. Bacon adopted the term only gradually, but by 1620 he used it frequently to refer to the phenomenon which he believed that he had been observing and describing for many years. In Bacon's reading of the Scriptures, he was on the cusp of an age of the world in which human knowledge was going to see tremendous advances, according to the divine plan. Thus, according to his interpretation of Daniel 12:4 in 1620:

> Nor should the prophecy of Daniel be forgotten, touching the last ages of the world:– "Many shall go to and fro, and knowledge shall be increased;" clearly intimating that the thorough passage of the world (which now by so many distant voyages seems to be accomplished or in course of accomplishment), and the advancement of the sciences, are destined by fate, that is by Divine Providence, to meet in the same age.[4]

If the event of the Instauration was decreed by God, it was also the result of human will and effort, for it is the humans themselves who go "to and fro" for the increase of knowledge. It would also be human beings who would, in the last ages of the world, actively read the book of nature through the process of observation and experiment. The result of this combination of divine and human action would be the Instauration, an age which would "restore (*instaurare*) and extend the power and dominion of the human race itself over the universe."[5] If human power and dominion is being restored, then it follows that there was a previous period when the human race had power and dominion over the universe. Many scholars have recognized Bacon's belief that, in reforming the study of natural philosophy, he was recovering the mastery over nature which humankind had once possessed in the Garden of Eden. This is certainly true, but there is more to it. Eden and the Instauration are but two elements, or *loci*, in a well-developed theological system embedded in Bacon's writings. The entire system can be seen in considering the passages in which Bacon retells the Christian narrative of "sacred history."

Bacon regarded "sacred history" as one of the three branches of the academic discipline of theology. In discussing what was necessary for theologians to contribute to the new learning in the *De Augmentis Scientiarum*, he separated the main heading

---

[2]   Ibid., p. 379.

[3]   Ibid., p. 377. The Greek ἀνακεφαλαιώσασθαι is better translated as "recapitulated" than "summed up," but the latter is found in the English translation of Irenaeus in the ANF.

[4]   WFB, vol. IV, pp. 91–2. Spedding's translation.

[5]   *Novum Organum*, Book 1, Aphorism 129; WFB, vol. I, p. 222. I have corrected the Spedding translation to properly reflect the Latin meaning of *instaurare*. Spedding translated it as "*establish* and extend.". See WFB, vol. IV, p. 114.

of theology into three approaches, none of which is original, but all of which reflect his observation of the field. Bacon made the approaches correspond to the three aspects of the human soul which he considered to be the natural and observable state of things – memory, reason, and imagination – and thus the approaches to theology already in place are doubly justified. Corresponding to reason is dogmatic theology, which is the study of divinity in the philosophical abstract. Corresponding to imagination are the parables, which are divine fables or poetry, the "literature" of divinity. Corresponding to memory is sacred history, which is the record of God's actions in the world of time and place. Unlike typical human history, sacred history includes prophecy, for this also is a record of God's actions in time and place, and in divine history "the narration may be before the event as well as after."[6]

Apart from certain passages in his early *Meditationes Sacrae*, Bacon very rarely approached theology from the abstract: he leaves dogmatic theology to the theologians. Nor does he concern himself much with constructing parables, unless, after we have considered the theology behind the Instauration, we might contend that he has done exactly that in his *New Atlantis*.[7] Instead, in the bulk of his writing, he approached the theology of the Instauration through sacred history. Christian theology operates in connection with a chronological narrative stretching from the creation, through the fall and the incarnation of Christ, ending in the return of Christ, and the new heaven and new earth. The whole body of Christian dogma can be presented in connection with the events of this narrative. This is the *historia sacra*, the *Heilsgeschichte*, of classical theology. When viewed specifically in light of the edenic fall and the means of recovery, it is also termed, "salvation history," for it tells the tale of the providential hand of God working for human recovery, and it reveals the *via salutis,* the progressive way in which that recovery is accomplished in time, both in respect to the Church and the individual. Throughout Bacon's Instauration writings he refers to this grand narrative of Christian theology, and he has placed the divine action of the Instauration event within this narrative, to function as an organic part of it. However, not only did he need to shift some standard furniture in order to make room for his new addition, but he also needed to give some sort of explanation in order to justify adding an element to the sacred narrative which had not been there before, and which only he had fully recognized for what it was.

The omission of the Instauration by theologians before Bacon is in keeping with his criticisms of the state of theology in his day as he set them forth in the *De Augmentis Scientiarum*. Although he generally avoided direct criticism in this work, he was quick to claim that theologians had neglected two crucial elements of sacred history: the "History according to Prophecy," and the "History of divine judgment or Providence." By the "History according to Prophecy" Bacon meant the activity

---

6   Translation from WFB, vol. IV, p. 293. This discussion was originally set forth in a much shorter form in *The Advancement of Learning*, which is the original version of the *De Augmentis Scientiarum.* Cf. WFB, vol. III, pp. 340–42.

7   Stephen McKnight presents this work as a religious allegory which supports the idea of the Instauration as an act of divine providence. In this light it would truly be a parable of the Instauration. See Stephen McKnight, *The Religious Foundations of Francis Bacon's Thought* (Columbia, 2006), pp. 10–44.

of connecting prophecies with their evident fulfillment in time. This had not been adequately done before Bacon, and when it was done in the future it had to be done with "great wisdom, sobriety, and reverence, or not at all."[8] It was essential that, when the task was done, prophecies should not be improperly limited in interpretation, as if it were a matter of strict one-to-one correspondence between prophecies and discrete events: "and though the height or fullness of them is commonly referred to some one age or particular period, yet they have at the same time certain gradations and processes of accomplishment through divers ages of the world."[9] By the "History of Divine Judgment or Providence" Bacon meant the manifestation of the will of God in the world, specifically through the coincidences of events which made it clear that the will of God was actively at work, for judgment, unexpected deliverance, or even the case of "divine counsels, through tortuous labyrinths and by vast circuits, at length manifestly accomplishing themselves."[10]

Both of these omissions on the part of theologians tie in with Bacon's retelling of the narrative of salvation history. On the one hand, as the earlier quotation from the *Novum Organum* attests, Bacon saw the Instauration as the fulfillment of specific prophecies which had not, up to his day, been properly understood. According to his discussion in *De Augmentis*, the history according to prophecy had simply not been adequately developed. On the other hand, as is also evident from the quotation from the *Novum Organum*, Bacon saw the Instauration as a fortuitous convergence of the expansion of the known world through discovery and the advancement of the sciences. The providential hand of God was clearly at work, though others had not recognized what was happening. According to the discussion of theology in *De Augmentis*, this could be ascribed to the deficient state of the history of providence among theologians, part of which discipline was the observance of "divine counsels" which "through tortuous labyrinths and by vast circuits" at last came to pass and could be observed. Bacon had observed what the theologians had not.

Throughout his writings dealing with the Instauration, Bacon amended the standard narrative of sacred history, weaving the Instauration into his exegesis of the Scriptures and his presentation of Church history. Many of these moves were only possible because of his theological shift away from Calvinism. In addition, Bacon's retelling of the narrative relied on the increased fluidity in interpretation of the Old Testament occasioned by the recovery of Hebrew, for he reinterpreted many passages, such as the prophecy of Daniel in the quotation earlier. Bacon did not know Hebrew, and he would have been a poor judge of the latitude that would have been acceptable among theologians of his day, but Lancelot Andrewes was a recognized authority in Hebrew. Andrewes was also Bacon's "inquisitor" on the text of the *Advancement of Learning* in which Bacon's amended narrative of sacred history was first published. It is doubtful whether Andrewes recommended many substantive changes. With only a few exceptions, the points made by Bacon, even in his unpublished writing, *Valerius Terminus,* are entirely in line with Andrewes' lectures on God and nature in the Αποσπασματια *Sacra.* Although parts of Bacon's

---

8   Translation, WFB, vol. IV, p. 313.
9   Ibid.
10   Ibid.

reading of Scripture are unique, they are nevertheless very close to the thinking of a man who, after 1605, was a respected bishop of the Church of England.

Although elements of the narrative of sacred history can be found throughout Bacon's philosophical writings, the narrative is most clearly laid out in a series of four works which were designed to provide a grand overview of the plan for the reform of learning. The first of these, chronologically, is the manuscript, *Valerius Terminus*, which Bacon seems to have abandoned in 1603, but apparently kept for reference as he wrote later works.[11] This work is Bacon's first draft of a publication designed to provide a description of the Instauration event, along with the implications of the new method for the study of nature he was proposing. Although it appears that the *Valerius Terminus* was superseded by his publication of *The Advancement of Learning* in 1605, the manuscript is still significant as it represents Bacon's earlier and most candid presentation of the material. Many of the arguments laid out in the *Valerius Terminus* are reproduced with very little modification in his later writings. *Two Bookes of the Proficience and Advancement of Learning* was published in 1605 and it represents Bacon's most thorough description of his project up to that point. In 1620 Bacon published the *Instauratio Magna*, along with the *Novum Organum*, in a single volume. Unfortunately, this work generally receives the greatest attention when Bacon's reform of learning is considered, for much of it is merely a summary of points made in *The Advancement of Learning*, and often in a simplified form, as the *Novum Organum* was designed to present the argument "digested into aphorisms."[12] In 1623 Bacon would entirely rework *The Advancement of Learning* with assistance from a select group of editors, and it would be republished in a much expanded form in Latin, as the *De Augmentis Scientiarum*. This work represents the most detailed discussion of Bacon's entire program for the reform of human learning. Bacon published many other works in connection with his grand project, such as the *Cogitata et Visa*, and the *Sylva Sylvarum*. These works are not to be ignored, but the clearest exposition of the rationale of the project, and the theology behind it, is found in the four works which were designed to give a broad overview of the Instauration. Although there were bound to be changes in these various works as Bacon modified and adapted his presentation, the theology changed very little over time.

The Ages of the World and the Chain of Causes

The beginning of the narrative of sacred history is always found in Genesis, the book of "beginnings." From the Genesis narrative of creation Bacon derived principles which were essential to the Instauration. The first principle was that Bacon's God was the God of order who had constructed an orderly and predictable universe. This principle had significant implications for the Instauration at a number of levels: Chronologically, it meant that the entire history of the cosmos could be divided into four discrete ages; cosmologically, it meant that the universe was so structured that it always operated according to rules which governed the realm of secondary causes;

---

[11]   See the observation on Harleian Ms. 6462 by Spedding in WFB, vol. III, p. 206.

[12]   "*sed tantum digestam per summas, in Aphorismos.*" WFB, vol. I, p. 146.

and, practically, it meant that if the human role in the Instauration were to be carried out properly, man must follow the prescriptive hierarchy which was manifested in the order in which all things came to be. We will consider these aspects in turn.

According to Bacon's *Confession of Faith,* God, through the mediation of the Logos, created heaven and earth to operate according to "constant and everlasting laws, which we call *Nature,* which is nothing but the law of the creation; which laws nevertheless have had three changes of times, and are to have a fourth and last."[13] The first period of time was when God had made matter, but had not yet begun the six days of creation when He would give it form. The second period includes the days of creation and the Sabbath, and it culminates in the fall. After the fall is the period of the curse, "which notwithstanding was no new creation, but a privation of part of the virtue of the first creation." There was a change in the laws of creation as a result of the fall in which the laws received a "revocation in part by the curse, since which time they change not." This did not involve a complete reordering of nature, but only a certain modification, and otherwise "the laws of Nature, which now remain and govern inviolably till the end of the world, began to be in force when God first rested from his works and ceased to create." It was at this point, after the direct act of forming matter in the course of the six days, that God's rule over nature proceeded "by compass," or a "chain of causes" as in the *Meditationes Sacrae,* rather than being "immediate and direct." There is also a fourth and final period, which will begin at the end of this present world. The scheme of the history of God and creation outlined in the *Confession* remained firm throughout Bacon's life. All of Bacon's statements regarding God and creation in the Instauration corpus relate to this basic chronological framework.

The concept of a "chain of causes" which we examined in connection with the *Meditationes Sacrae* is an important part of the discussion of nature in *The Advancement of Learning,* where its implications for the reform of human learning can be seen. Bacon responded specifically to the objection that the pursuit of earthly knowledge led to atheism:

> And as for the conceit that too much knowledge should incline a man to atheism, and that ignorance of second causes should make a more devout dependence upon God which is the first cause; first, it is good to ask the question which Job asked of his friends, *Will you lie for God, as one man will do for another, to gratify him?* For certain it is that God worketh nothing in nature but by second causes; and if they would have it otherwise believed, it is mere imposture, as it were in favour towards God; and nothing else but to offer the author of truth the unclean sacrifice of a lie.[14]

The target here is not only some group that objected to the *knowledge* of second causes, but also, as the last sentence makes clear, those who *denied* second causes or the chain of causation completely because, according to them, nothing happened in the world save by the immediate will and operation of God. Bacon was arguing against those who took Calvin's argument in its most literal sense. In his hyperbole he accuses them of lying. He is countering, in advance, the suggestion that *he* was

---

[13]   All of the following quotations are from WFB, vol. VII, p. 221.

[14]   WFB, vol. III, p. 267.

being impious by accusing those who deny second causes of impiety themselves. In order to preserve God's power they were denying what Bacon regarded as the obvious order of the cosmos as God had set it up. In denying the obvious under the pretense of piety, they *were* lying after a fashion, even if they were only lying to themselves. For Bacon, nature followed an ordinary course in obedience to the laws which God established in creation, but this was not to be attributed to His immediate action as Calvin would have it. Therefore, in the same vein as in *The Advancement of Learning,* Bacon stated in the *Valerius Terminus*: "That a religion ... that cherisheth devotion upon simplicity and ignorance, as ascribing ordinary effects to the immediate working of God, is averse to knowledge."[15]

The doctrine of God acting through a chain of causes, rather than immediately, was critical for Bacon's understanding of the Instauration. As we have already noted, if God is the immediate cause of everything, there is no room for genuine free will in man, and God is necessarily the author of evil. In Bacon's understanding, experiments were essential to bring about the reform of learning, but, if God always acts immediately, experimentation is not possible. God is then the only true agent, and nature cannot, in any meaningful sense, be freely manipulated by humans.[16] The chain of causes is a space in which the experimental method can operate, and in which humankind, as free willing agents, can achieve some supremacy and control over other causes.

A God who acts immediately in nature must be the cause of evil. For Bacon, evil had its origin farther down on the chain of causes, as a "perversion" of that which was created good. The first problem arose with the rebellion of humanity, but the rebellion didn't stop with the fall of humankind alone. Bacon regarded the fall as an event which disrupted the entire cosmos, and damaged the chain of causes itself. Among the consequences of the fall, nature had also entered into a state of rebellion or waywardness. In the *Valerius Terminus,* Bacon described the difficulty of humanity regaining mastery over nature as the result of nature being "turned to reluctation."[17] This is how he accounted for the suffering, misery, and hardship which afflicted human beings after the fall. The chain of causes not only allowed for human agency, but also for agency in nature apart from the immediate action of God or the influence of humans. In *De Augmentis Scientiarum,* Bacon described nature, when it is not operating according to the law of God in creation or the law imposed by man through mechanical arts, as "driven out of her ordinary course by the perverseness, insolence, and frowardness of matter."[18] Nature might still be found following the course set by God, but, after the fall, matter had about it a "perverseness," which can cause it to veer from that course. Humankind can choose, through art, to influence

---

[15]   Ibid., p. 251.

[16]   This is what Calvin maintained in *Institutes*, Book I, ch. 16, 1–3. The human invention of arts is not to be attributed to human agency, but to the actions of God through His human instruments, according to *Institutes*, Book I, ch. 5, 5, in which the "inventing of so many wonderful arts are sure indications of the agency of God in man." See John Calvin, *Institutes of the Christian Religion*, trans. Henry Beveridge (Edinburgh, 1845).

[17]   WFB, vol. III, p. 222.

[18]   WFB vol. IV, p. 294. Original Latin in WFB, vol. I, p. 496.

and control nature, which is what Bacon is striving for in his reform of learning. But if this is to be done correctly, Bacon contended, it requires the imitation of God's own order and method in creation.

Creation as a Pattern for Human Learning

From the Genesis creation narrative Bacon repeatedly drew attention to the fact that God had created light first, and argued that this should inform human efforts in experimentation and natural philosophy. In Aphorism 70 of the *Novum Organum* Bacon expressed it this way:

> But in the true course of experience, and in carrying it on to the effecting of new works, the divine wisdom and order must be our pattern. Now God on the first day of creation created light only, giving to that work an entire day, in which no material substance was created. So must we likewise from experience of every kind first endeavor to discover true causes and axioms; and seek for experiments of Light, not for experiments of Fruit.[19]

If the matter of creation is to be manipulated by human activity it is essential, according to Bacon, that we bear in mind the way in which the universe has been assembled. This requires not only examining the fabric of the universe to understand how its various elements fit together (which was essentially reading the Book of Nature), but also remembering that God, who has done all things the right way, has revealed the proper order for such a project in the chronological arrangement of the events of creation itself. Even as God created light first, and devoted an entire day to the activity, so mankind must proceed slowly and first come to understand the creation before attempting to control it and make it produce. This is more than a convenient analogy.

In *The Advancement of Learning* Bacon identified the habit of a "peremptory reduction of knowledge into arts and methods"[20] as an error of previous generations, which had to be remedied by looking to the revealed actions of God, who is the "arch-type" of knowledge:

> First, therefore, let us seek the dignity of knowledge in the arch-type or first platform, which is in the attributes and acts of God, as far as they are revealed to man and may be observed with sobriety; wherein we may not seek it by the name of learning; for all learning is knowledge acquired, and all knowledge in God is original: and therefore we must look for it by another name, that of wisdom or sapience, as the Scriptures call it.[21]

Here, Bacon defines "learning" as the human acquisition of God's wisdom. This divine wisdom, the pattern for human "learning," was expressed first in the act of creation itself, which proceeded according to a governing principle of hierarchy. In commenting on the hierarchies of angels listed by Pseudo-Dionysius the Areopagite, Bacon observed that God always acted in a hierarchical pattern which had love

---

[19]   WFB, vol. I, p. 180. Translation in WFB, vol. IV. p. 71. See also the same argument differently worded in the preface to the *Instauratio Magna*, WFB, vol. I, pp. 128–9.

[20]   WFB, vol. III, p. 292.

[21]   Ibid., p. 295.

(the ultimate motivation) at the top, followed by "light," which Bacon equated with knowledge.[22] In the pattern of creation both these causes are antecedent to, and thus necessary for, the manipulation of nature by art. Bacon regarded the order in which God operated, placing light before production and spending a significant amount of time in creating light, as normative for human method precisely because he regarded divine wisdom as the archetype for human learning.

Up to this point the similarity between Bacon's statements and the content of Andrewes' sermons on Genesis 1–4 is striking. For Andrewes, the act of creation, every bit as much as the act of redemption, was the work of Christ as the "mediator" between God and creation.[23] While the *Αποσπασματια Sacra* does not go into the detail of the *Confession of Faith*, the Logos theology is unmistakeable. After creating matter, God proceeded to give it form according to an orderly pattern. The purpose of God's order, and particularly His revelation of it to us through Moses, was not only to present Himself as the God of order, but to guide us in our own meditation on nature: "God also took this orderly proceeding, partly that we entering into the meditation of God's works, might have, as it were, a thread to direct us orderly therein."[24] For Andrewes, as for Bacon, the order of creation itself was a pattern for human consideration of nature, or "natural philosophy." Light was, according to Andrewes, quite naturally the first of God's creatures, for it is the creature by which all other things are distinguished from one another, and it is directly related to the human faculty of knowledge: "for all our knowledge cometh of light, and is compared to light."[25] Significantly, the activity of giving form to matter is presented by Andrewes as an act of distinction and dividing, increasing by increments the complexity of created order from the first day in which God distinguished light from darkness.[26] As part of the process of this act of continual distinction, "in all the six dayes works, God gave names to the things as he made them, and to Adam himself, and in these seven things named, are contained all other particular things made in, and with them."[27] The act of divine naming had a specific purpose for man: "as God gave … the natural use of things, so now he took order that we might have a use of them by names, to know and talke of them so."[28] This brings us to Adam's place and activity in the Garden.

Humanity in the Garden

As we have already observed in Bacon's *Confession of Faith*, human beings were created to have a special place within the hierarchies of creation as the creatures to whom God would unite His own divine nature in the hypostatic union at the incarnation. From the beginning, humanity was designated as the point of contact through which

[22] Ibid., p. 296.
[23] Lancelot Andrewes, *Αποσπασματια Sacra* (London, 1657), p. 38.
[24] Ibid., p. 11.
[25] Ibid., p. 23. See also p. 55.
[26] Ibid., pp. 25–6.
[27] Ibid., p. 33.
[28] Ibid.

God would establish a communication between Himself and the entire cosmos. Even before the incarnation, however, humans held a special place of rulership over material creation, though not yet over the angels. According to the *Confession of Faith*, "God created Man in his own image, in a reasonable soul, in innocency, in free will, and in sovereignty."[29] Bacon is not unique in asserting humankind's sovereignty over the lower orders of creatures. Human sovereignty over the earth is a theological commonplace based on God's command to humanity to "subdue the earth" in Genesis 1:28. The key question is how this sovereignty is exercised.

In the *Valerius Terminus* Bacon declared that God had created the human mind for the purpose of investigating and understanding the universe:

> God hath framed the mind of man as a glass capable of the image of the universal world, joyning to receive the signature thereof as the eye is of light, yea not only satisfied in beholding the variety of things and vicissitude of times, but raised also to find out and discern those ordinances and decrees which throughout all these changes are infallibly observed.[30]

Humans were designed to investigate and learn the laws of nature which God has laid out for the governance of creation. Bacon supported this point with a reference to Proverbs 25:2, a verse which would also figure prominently in his preface of the *Instauratio Magna*: "It is the glory of God to conceal a thing, but it is the glory of the King to find it out." In his explanation of this verse in *Valerius Terminus* Bacon does not allow it to refer specifically to kings, but to the general vocation of humankind from the beginning:

> Nay, the same Salomon the king affirmeth directly that the glory of God *is to conceal a thing, but the glory of the king is to find it out*, as if according to the innocent play of children the divine Majesty took delight to hide his works, to the end to have them found out; for in naming the king he intendeth man, taking such a condition of man as hath most excellency and greatest commandment of wits and means, alluding also to his own person, being truly one of those clearest burning lamps, whereof himself speaketh in another place, when he saith *The spirit of man is as the lamp of God, wherewith he searcheth all inwardness*.[31]

In *The Advancement of Learning* Bacon presented the work of Adam in the Garden as an act of learning:

> After the creation was finished, it is set down unto us that man was placed in the garden to work therein; which work so appointed to him could be no other than the work of contemplation; that is, when the end of work is but for exercise and experiment, not for necessity; for there being then no reluctation of the creature, nor sweat of the brow, man's employment must of consequence have been a matter of delight in the experiment, and not matter of labour for the use. Again, the first act which man performed in Paradise consisted of the two summary parts of knowledge: the view of creatures, and the imposition of names.[32]

---

[29]   WFB, vol. VII, p. 221.

[30]   WFB, vol. III, p. 220.

[31]   Ibid.

[32]   Ibid., p. 296.

The activity of Adam in Eden was that of the natural philosopher. He contemplated nature and actively experimented with it in order to experience the joy of understanding the universe which God had made, and seeing what could be done with it. The work of Adam in Eden was the same as that work which Bacon was proposing in the Instauration writings, but with the significant difference that Adam did not have to contend with the rebellion, or "reluctation," of a natural world corrupted by sin.

By engaging in this work of the contemplation of and experimentation with nature, Adam was partaking in a truly divine activity, for the high point of the creation narrative was the seventh day when God contemplated his works: "So in the distribution of days, we see the day wherein God did rest and contemplate his own works, was blessed above all the days wherein he did effect and accomplish them."[33] The activity of Adam in the Garden of Eden, and of his children in the instauration, is closely tied to the identity of humans in that it was made in the very image of God. Adam and his Maker were united in the contemplation of creation. Adam, the image of God, learned through investigation and experimentation what God knew about His own power in the cosmos. This understanding of the human vocation in perfection is entirely in keeping with the doctrine of the Cappadocian Fathers, particularly Basil the Great and Gregory of Nyssa, who were widely read in the late sixteenth and early seventeenth centuries. Gregory of Nyssa summarized the place of humanity in the cosmos as one of observation and learning, beholding, through the things of creation, "that power of the Maker which is beyond speech and language."[34]

Adam's act of the observing and naming of the creatures (Gen. 2:19–20) was nothing other than observing the creatures and identifying them according to their roles, functions, and uses in the divinely established hierarchies. In the perfect knowledge of created things which existed in Adam before the fall, Bacon says, he "did give names unto other creatures in Paradise, as they were brought before him, according unto their properties."[35] The act of naming was not merely a nouthetic activity, but was integrally associated with human power over the lesser creatures. Humans were not only to identify the earth and appreciate it, but to subdue it, expressing their identity as God's image in creation by exercising the power which God gave them over the lower orders in the hierarchy. The connection of naming to human power over creation has significant overtones of Renaissance magic which become unmistakeable in Bacon's claim in *Valerius Terminus* that "whensoever he [man] shall be able to call the creatures by their true names he shall again command them."[36] However, we must remember that what is implied here, magical though it may be in its derivation, is more sophisticated than basic incantation.

For Bacon, the "name" is always the identification of the thing according to its true function and use. Naming is recognizing the properties of a thing and its place in creation so that it may be used according to its properties. Adam's activity of naming

---

[33] WFB, vol. III, p. 296.

[34] Gregory of Nyssa, *On the Making of Man*, NPNF, series 2, vol. 5, p. 390.

[35] WFB, vol. III, pp. 264–5.

[36] WFB, vol. III, p. 222. On the role of names in magic see D.P. Walker's discussion of the *Vis Verborum* in his *Spiritual and Demonic Magic from Ficino to Campanella* (London, 1958), pp. 80–81.

was a matter of discriminating between creatures, and defining them, according to their properties. The edenic naming of the creatures, then, is akin to Bacon's doctrine of "forms," or the "formal cause" as Bacon has adopted and adapted the term from Aristotelian natural philosophy. In the second book of the *Novum Organum*, Bacon explained the meaning of "Forms" in his system as follows:

> For though in nature nothing really exists beside individual bodies, performing pure individual acts according to a fixed law, yet in philosophy this very law and the investigation, discovery, and explanation of it, is the foundation as well of knowledge as of operation. And it is this law, with its clauses, that I mean when I speak of *Forms*; a name which I the [*sic*] rather adopt because it has grown into use and become familiar.[37]

This definition of "Forms" is more developed than his parallel discussion in the *Valerius Terminus*, where they are simply left as the "true differences" between the things of nature, but it cannot be doubted that the concept was already present in seminal form in the earlier work.[38] The discovery of Forms in both the *Novum Organum* and the *Valerius Terminus* is characterized by the actions of "dividing" and "defining," of which Plato despaired, saying that "he will revere him as a God, that can truly divide and define."[39] In a fallen world, many steps are now required for humanity to divide and define nature, and recover the Forms by which the specific bodies of nature operate, including the direct observation of natural bodies and motions, and the gathering of specific instances. Adam, for his part, was able to perform this action of division and definition merely upon seeing the creatures before him.

Again, Lancelot Andrewes has a parallel discussion in his consideration of the human act of "naming":

> All names man giveth is of the property; we say commonly that this is the nature, *scilicet*, the propertie of a thing: The knowledge of which properties is either sensible of outward things, or intelligible of inward qualities. The names of things after *Adam* were of properties sensible, as *Esau* was so called, for that he was red and rough with haire … But *Adams* names came from inward qualities, which he could perceive partly by the light of nature.[40]

Andrewes regarded false knowledge as a misunderstanding of the place of creatures in creation, and a failure to recognize God's intent for them, their properties, and their true natures:

> The end, to which God gave & imposed sundry names was, that we should do as he hath done, that is, when things have a true being, then to give names to them accordingly, and not to our fancies. ... for as man draweth good Liquor out of the Cask, so out of the

---

[37]   Aphorism II, of the *Novum Organum*, WFB, vol. I, p. 228. Translation, WFB, vol. IV, p. 120.

[38]   *Valerius Terminus*, WFB, vol. III, p. 239.

[39]   For the use of this quotation in *Valerius Terminus* see WFB, vol. III, p. 239. For the *Novum Organum* see WFB, vol. I, p. 277.

[40]   Andrewes, Αποσπασματια *Sacra*, pp. 213–14.

meaning of the Word, and denominations given by God, we may draw out the hidden nature and knowledge of the thing.[41]

If we call things by their correct names, then we are identifying them according to both the properties and the use which God gave them in creation; otherwise, we are using names improperly. Thus, in the introduction to his *Historie Naturall and Experimentall*, Bacon also expresses his frustration that his own generation is still averse to true knowledge: "For we create worlds, we direct and domineer over nature, we will have it that all things are as in our folly we think they should be, not as seems fittest to the Divine wisdom, or as they are found to be in fact."[42] The error of his generation was in refusing to learn God's intent for nature, and recognize things according to their true "names."

For Andrewes, Adam's activity in the Garden did not stop with the act of naming, or the observing of the properties and use of created things. It proceeded on to experiment: using all things according to their properties. For "God made the Earth as his work-house and shop, and Heaven as his chamber and place for a rest and reward, and both for one; and that is man."[43] Adam was placed specifically in the workshop of the Garden in order to manipulate it, and learn how to make it produce:

> But all the Fathers doe agree in this, that it was Gods will that the Garden should bring forth, not only *opera naturalis*, of his own accord, but also by the industry and diligence of man, it should bring *opus voluntarium*. So that divers other faire and pleasant things should be bestowed on the Garden, and caused to grow by his labor, and so he should both *discere & docere*, how many things by industry might be done above nature.[44]

As Bacon saw Adam's activity as experimentation to discover the potential of all nature, so Andrewes presented Adam as engaged in the business of the laboratory: a "hands-on" program of learning and demonstration (*discere et docere*) of what could be done with the material of nature that was at hand. For both Bacon and Andrewes the fall of humanity into sin changed everything. The key question was – as always – how.

Knowledge and the Fall

In arguing that human knowledge could and should be advanced and developed, Bacon had to contend with the role that knowledge may or may not have played in the fall itself, for the first sin was nothing other that eating from a tree of knowledge. In this area, Bacon was setting himself against the opinions of many prominent theologians of his day. Nor did he step very carefully, particularly once he had risen to power. In 1605, in *The Advancement of Learning,* Bacon was concerned with refuting a specific objection of certain "divines" who held:

---

[41] Ibid., p. 33.

[42] Spedding's translation of Bacon's introduction to *Historia Naturalis et Experimentalis* of 1622, WFB, vol. V, p. 132. Cf. the Latin in WFB, vol. II, p. 14.

[43] Andrewes, Αποσπασματια *Sacra*, p. 121.

[44] Ibid., pp. 179–80.

... that knowledge is of those things which are to be accepted of with great limitation and caution; that the aspiring to *over*-much knowledge was the original temptation and sin, whereupon ensued the fall of man; that knowledge hath in it somewhat of the serpent, and therefore when it entereth into a man it makes him swell.[45]

In 1620 Bacon provided a more detailed image of the reasoning of his opponents, and a harsher censure of them, in the *Novum Organum*:

Lastly, you will find that by the simpleness of certain divines, access to any philosophy, however pure, is well nigh closed. Some are weakly afraid lest a deeper search into nature should transgress the permitted limits of sobermindedness; wrongfully wresting and transferring what is said in holy writ against those who pry into sacred mysteries, to the hidden things of nature, which are barred by no prohibition. Others with more subtlety surmise and reflect that if second causes are unknown everything can more readily be referred to the divine hand and rod; a point in which they think religion greatly concerned; which is in fact nothing else but to seek to gratify God with a lie.[46]

The divines of Bacon's day were primarily Calvinist. It was a Calvinist understanding of human knowledge, or a Calvinist caution about it, which is the target of these statements.

In Calvin's discussion of the fall, human knowledge was heavily implicated. Eve's error, according to Calvin's lecture on Genesis 3, lay in trusting her senses, which told her that the tree was good for food, and in desiring greater knowledge of the things of the Garden than she was permitted to have:

But [Eve] erred, whiche tempered not the measure of knowledge with the will of God. And we all daily sicke of the same disease, in that we desire to know more than is meete and than the Lord permitteth; seeing the principall point of wisdom is, framed sobriety to the obedience of God.[47]

Eve had entered into an unlawful study of the created tree, and her observation led her to, in Bacon's words, "the aspiring to *over*-much knowledge." This was the position to which Bacon was objecting in the passages from *Valerius Terminus* and *The Advancement of Learning*. Knowledge, and trust in the senses, was set in opposition to faith in Calvin's discussion. Eve had turned from true knowledge provided by revelation to the deceitful knowledge provided by the senses. According to the *Institutes*, it was the desire for the extra knowledge the tree could provide which was sinful because "man" was "seeking more than was granted him."[48] It does not follow, of course, that human knowledge was completely proscribed for Calvin or for later Calvinists, many of whom were heavily engaged in the scientific activities of the Royal Society and its predecessor institutions.[49] However, there is a

---

[45]   WFB, vol. III, p. 264.

[46]   WFB, vol. I, p. 197. Translation, WFB, vol. IV, p. 88.

[47]   *A Commentarie of John Caluine, upon the first booke of Moses called Genesis*, trans. Thomas Tymme (London, 1578), Chapter 3, sec. 5. p. 91.

[48]   ICR, Book 2, ch. 1, sec. 4. Battles translation, p. 245.

[49]   See Charles Webster, *The Great Instauration: Science, Medicine, and Reform 1626–1660* (New York, 1975).

serious difference between Calvin and Bacon on the understanding of the fall here, and the disparagement of the pursuit of human knowledge was a common theme of some of the most prominent Calvinists of Bacon's day.

William Perkins was a staunch Calvinist and an influential teacher at Cambridge who personally resisted Whitgift's pressure for conformity. According to Perkins' *Discourse of the Damned Art of Witchcraft*, the desire for knowledge of nature which God had not granted was one of the dangerous "discontentments" of the mind which resulted in the practice of witchcraft.[50] There were many thousands of things that man, according to his natural limitations, could never know in the natural world, but Satan accomplished wonders by his tremendous knowledge of natural things.[51] In desiring the secret and forbidden knowledge wielded by the devil, humans were following a dangerous example, and the result would ensnare the human soul in satanic intrigues.

Another significant example of the effect of Calvin's doctrine on English thought is found in Thomas Cartwright, Archbishop Whitgift's perennial Puritan target, whose work Bacon read while at Gray's Inn.[52] In his discussion of the fall in *A Treatise of Christian Religion,* Cartwright paralleled the exact argument of Calvin's Genesis commentary in presenting the sin of Eve as the desire for forbidden knowledge, which came through the tree. He elaborates that those things which God has not revealed should not be sought, and that "ignorance in such things, is the best knowledge."[53] Cartwright also discussed the fact that Eve's temptation to the forbidden knowledge came through the senses of sight and hearing, both of which she trusted in believing the fruit good to eat. From this, Cartwright concluded that knowledge derived from the outward senses could be particularly deceptive now, after the fall, and that these senses should be at all times guarded: "they are (as it were) windowes, whereby sinne entered into the heart when there was no sinne; and therefore will much more now, the heart being corrupted."[54]

Throughout the Instauration writings, Bacon took great care to ensure that human knowledge was not implicated in the fall of Adam into sin, and argued consistently that the fall was the result of pride, rather than forbidden knowledge. If this position placed him at odds with certain Puritans from his youth, it was nevertheless squarely in line with the theology of Lancelot Andrewes. In several lectures dealing with the Tree of the Knowledge of Good and Evil and the condition of Adam before the fall, Andrewes argued that knowledge, being good and the creation of God, could in no way be held responsible for the fall. From the time of creation Adam was designed to know all that could intellectually be known. In regard to created things, Andrewes maintained

---

[50]   *A Discourse of the Damned art of Witchcraft*, in *The Workes of ... William Perkins* (3 vols, London, 1626–31), vol. 3, p. 609. I am indebted to Perez Zagorin for the observation that the *Discourse* presents a view of knowledge at odds with Bacon's. See Perez Zagorin, *Francis Bacon* (Princeton, 1998), p. 47.

[51]   Perkins, *Discourse*, pp. 610–11.

[52]   Lisa Jardine and Alan Stewart, *Hostage to Fortune: The Troubled Life of Francis Bacon* (New York, 1998), p. 79.

[53]   Thomas Cartwright, *A Treatise of Christian Religion, or, The Whole Bodie and Substance of Divinitie* (London, 1616), p. 49.

[54]   Ibid., p. 51.

that "Adam knew all things, not only perfectly, but exactly."[55] This knowledge of all things came to Adam naturally: "God gave him wisdom, he learned it not."[56] Only the essence of God and his hidden will were excluded from Adam's knowledge, but these transcend humanity's created reason and hence, by definition, are unknowable. All of the created cosmos fell within the grasp of man's knowledge. For Andrewes, this was because man was designed to be God's "viceregent,"[57] and "lieutenant,"[58] lord of all things save God himself, and knowledge was necessary for rulership. For Andrewes, Adam's perfect knowledge of all created things applied also to the forbidden tree.

Andrewes explained that Adam had complete understanding of the Tree of the Knowledge of Good and Evil, in that Adam *knew* all that there was to *know* about it:

> But some may say, *What hurt is it to know good and evill?*...
> I answer that God forbiddeth not to eat the fruit, nor that he would have us ignorant of that knowledge, *quam quis quaerit a Deo, sed quam quis quaerit a scipso*, And no doubt Adam had the knowledge both of good and evill, *per intelligentiam & si non per experientium*. And he knew how to choose the one, and to refuse the other, to pursue the one, and to fly from the other, he understood it then, but when he would know both by experience, Gen. 3.6. He could not see why God should forbid him, and therefore the Tempter taking occasion by it, made him make an experiment of it.[59]

According to Andrewes' discussion of Genesis 3:22, where God says "Behold, the Man is become as one of us, to know good and evil," the experience of rebellion did not really add to humanity's basic knowledge. Andrewes saw a measure of irony in this passage because the "knowledge" aspect of the tree did not come from eating of the tree, but from knowing what God said about it in His command. In knowing of the tree Adam and Eve knew that to obey God was good and disobedience was evil.[60] The serpent had merely confused the matter, and introduced an element of ignorance. In pride, Adam and Eve believed that there *was* something, which they deserved, being withheld from them. In Andrewes' opinion it is ironic that Satan's temptation offered Adam and Eve what they already had: they were already like God, being created in His image, and they already had the "knowledge of good and evil," though not the experience of it.[61] This may be profitably compared with Bacon's summary statement regarding the nature of the serpent's deception in *The Advancement of Learning*:

---

[55]   Andrewes, Αποσπασματια *Sacra*, p. 212.

[56]   Ibid.

[57]   Ibid., p. 96.

[58]   Ibid., p. 211.

[59]   Ibid., pp. 166, and 189, where the same point of experimental knowledge of good and evil is referred to St Augustine (though it is perhaps more Andrewes' interpretation than the intent of Augustine here).

[60]   Andrewes, Αποσπασματια *Sacra*, p. 189: "God, then by forbidding them to eat of the tree of knowledge, did not envy or grudge that they should have knowledge, but rather made this rule the root of all knowledge to them, that the science of good and evil is taken only from Gods *discendo*, that is, things are therefore good because God by his word alloweth them, and evill because he forbiddeth them."

[61]   Ibid., pp. 336, and 264.

As for the knowledge which induced the fall, it was, as touched on before, not the natural knowledge of creatures, but the moral knowledge of good and evil, wherein the supposition was, that God's commandments or prohibitions were not the originals of good and evil, but that they had other beginnings, which man aspired to know."[62]

It is a further irony that, now, the knowledge which Adam possessed in Eden by nature can only be recovered in the fallen world by experience and "making an experiment of it." But for Andrewes, as for Bacon, Adam's knowledge could, to some degree, be recovered.

For Andrewes, the first sin was a matter of the hubris of humanity eclipsing the wisdom which God had given to man. Bacon's discussions of the fall follow this reasoning exactly. Early on in the *Valerius Terminus* he raised the issue of the nature of the temptation which caused humanity to fall:

Man on the other side, when he was tempted before he fell, had offered unto him this suggestion, *that he should be like unto God*. But how? Not simply, but in this part, *knowing good and evil*. For being in creation invested with sovereignty of all inferior creatures, he was not needy of power or dominion; but again, being a spirit newly inclosed in a body of earth, he was fittest to be allured with appetite of light and liberty of knowledge; therefore this approaching and intruding into God's secrets and mysteries was rewarded with a further removing and estranging from God's presence.[63]

The fall was not occasioned by knowledge *per se*, and certainly not by the knowledge of nature, as some in Bacon's day believed. The fault lay elsewhere:

For it was not that pure and uncorrupted natural knowledge whereby Adam gave names to the creatures according to their property, which gave occasion to the fall. It was the ambitious and proud desire of moral knowledge to judge of good and evil, to the end that man may revolt from God and give laws to himself, which was the form and manner of the temptation.[64]

Adam and Eve wanted to revolt from God. The problem is not knowledge at all, but the sin lies in the selfishness and arrogance of the human motivation *for* the knowledge of good and evil. Adam and Eve ate of the tree in a desire to switch places with God and make their own rules, in order not to be dependent on God at all.

In the above quotation from *Valerius Terminus,* Bacon associated the sin with an intrusion into the "secrets and mysteries" of God. There is a common theme running through Bacon's writings that, in the act of sin, Adam and Eve were rejecting God's commandments or his "revelation." They knew that eating from the tree was wrong. God had revealed to them that this was not to be done, and that the eating bore serious consequences. In this sense the original sin, according to Bacon, was the result of a rejection of the knowledge concerning the tree that Adam and Eve already had, rather than an improper desire for further knowledge. In order to fully appreciate how this point operates in the course of Bacon's theology, we must recognize his

---

[62] WFB, vol. III, pp. 296–7.
[63] Ibid., p. 217.
[64] WFB, vol. I, p. 132; translation, WFB, vol. IV, p. 20.

assumption of the doctrine of the *Deus absconditus*, or the hiddenness of God. The *Deus absconditus* – the idea that there is always that in God which is simply beyond knowing – is a commonplace in Christian doctrine, stemming from the inherent differences between creatures, who are always finite or bounded, and God, who is unoriginate, transcendent, and infinite. However, Christian theologians are not always in agreement on where the line between the knowable and the unknowable should be drawn.

In the *Confession of Faith*, Bacon drew the line between the knowable and the unknowable precisely where Andrewes drew it, by distinguishing between the laws of nature, and the laws of God's hidden and secret will, which governed His interaction with spiritual creatures:

> At the first the soul of Man was not produced by heaven or earth, but was breathed immediately from God; so that the ways and proceedings of God with spirits are not included in Nature, that is, in the laws of heaven and earth; but are reserved to the law of his secret will and grace.[65]

The human mind was designed to comprehend nature or the universe, not the inner workings and plans of the transcendent God. God's personal interactions with "spirits," including the human soul, are beyond human comprehension, and this included such matters as God's giving of commandments to humankind for their benefit. God, being eternal and the designer of all, could comprehend the goodness of the Tree of the Knowledge of Good and Evil *being* in the Garden, as well as the goodness of His prohibition of the eating of that tree. Humanity, being time-bound and finite, could never comprehend the reasoning, or the perspective, of God, who is by nature eternal and transcendent. Nevertheless, God had directly revealed that part of His will which could be comprehended by humans – namely, that the tree was not to be used for food, and that there would be specific consequences if it was. In eating from the tree Adam and Eve rejected God's revelation and sought to know that which was, by definition, unknowable.

A corollary to this interpretation of the fall is that there is absolutely nothing in the investigation of the material universe which is either proscribed or necessarily beyond human comprehension. This idea will unfold more completely as we consider other aspects of Bacon's Instauration writings, but some mention should be made here of Bacon's oft-repeated discussion of the role of natural philosophy as a support for the faith, for this, too, is based on his line between the knowable and the *Deus absconditus*.

Knowledge as a Support for the Faith

In *The Advancement of Learning* Bacon wrote:

> It is an assured truth and a conclusion of experience, that a little or superficial knowledge of philosophy may incline the mind of man to atheism, but a farther proceeding therein doth bring the mind back again to religion; for in the entrance of philosophy, when the second causes, which are next unto the senses, do offer themselves to the mind of man,

---

[65]   WFB, vol. VII, p. 221.

if it dwell and stay there, it may induce some oblivion of the highest cause; but when a man masseth on farther, and seeth the dependence of causes and the works of Providence; then, according to the allegory of the poets, he will easily believe that the highest link of nature's chain must needs be tied to the foot of Jupiter's chair.[66]

If natural philosophy necessarily leads to a recognition of the divine, by ascending through the chain of causes, the caution must always be borne in mind that natural philosophy should not be thought to open into the *Deus absconditus*, and reveal that which is beyond comprehension. As Bacon said it a little earlier in *The Advancement of Learning*:

> If any man shall think by view and inquiry into these sensible and material things to attain that light whereby he may reveal unto himself the nature or will of God, then indeed is he spoiled by vain philosophy: for the contemplation of God's creatures and works produceth (having regard to the works and the creatures themselves) knowledge; but having regard to God, no perfect knowledge, but wonder, which is broken knowledge.[67]

Bacon is not suggesting here, as the last two phrases have sometimes been interpreted, that natural philosophy reveals nothing specifically about God and His work in the world. In context, he is stating that the things which are genuinely hidden from human reason – the nature and will of God – are inaccessible through natural philosophy. For Bacon, natural philosophy always reveals something about God: namely, his power manifested in a well-ordered universe. Later in the work, Bacon states that one of the essential functions of natural philosophy is "opening our belief, in drawing us into a due meditation of the omnipotency of God which is chiefly signed and engraven in his works."[68] Nor should it be inferred from passages such as those above that nothing can be known concerning the nature or will of God, for God has revealed much about both directly in the Scriptures. As Bacon had interpreted Matthew 22:29 in the *Meditationes Sacrae*, nature and the Scriptures were complementary theological sources, the former revealing God's power, and the latter his will. But he always makes a distinction between that which may be learned through observation and that which must be revealed.

Bacon and Andrewes have drawn the line between the knowable and the unknowable in God in much the same way as Irenaeus of Lyons. For Irenaeus, it was only through direct interaction with the second person of the Trinity, the Logos – or Christ – that anything could be known of what was otherwise part of God's hidden nature or secret will. Revelation was required, and that revelation was only made complete through the incarnation of Christ. Irenaeus wrote:

> There is therefore one God, who by the Word and the Wisdom created and arranged all things; but this is the Creator (Demiurge) who has granted this world to the human race, and who, as regards His greatness, is indeed unknown to all who have been made by Him (for no man has searched out His height, either among the ancients who have gone to their rest, or any of those who are now alive); but as regards His love, He is always known through Him by whose means He ordained all things. Now this is His Word, our Lord

---

[66]   WFB, vol. III, pp. 26–8.
[67]   Ibid., p. 267. See also the parallel section in *Valerius Terminus*, ibid., p. 218.
[68]   Ibid., p. 301.

Jesus Christ, who in the last times was made a man among men, that he might join the end to the beginning, that is man to God.[69]

What was revealed through creation itself (by the action of the Logos, again) was nothing more or other than God as creator. This is what Bacon regarded as the knowledge of God's power, and certainly the manifestation of God according to His creation is clear to all who can see that creation, according to Irenaeus:

For by means of the creation itself, the Word reveals God the Creator; and by means of the world [does He declare] the Lord the Maker of the world; and by means of the formation [of man] the Artificer who formed him; and by the Son that Father who begat the Son.[70]

It was only in the incarnation that anything could be known of God beyond His power and majesty as Creator. It was only through the visible Son that the Father could be personally known. Thus, according to Irenaeus, Moses, the prophets, and all who came before the incarnate Christ could know nothing of God's nature or will which had not been directly revealed by an interaction with the then pre-incarnate Christ. This occurred only through specific theophanies, as in Isaiah's heavenly vision or the burning bush of Moses. Irenaeus noted in the case of Elijah, in 1 Kings 19:11–12, that God made it clear that He was not to be found in the powerful wind, in the earthquake, or in a fire. Only in the "scarcely audible voice" [*vox aurae tenuis*] of his revelation, did Elijah have any knowledge of God's will or His plan for man.[71] Indeed, God's power could be known through the things of creation themselves, but anything of His will or His nature can only be known through the direct action of Christ in the world.

The distinction evident in this discussion is operating throughout Irenaeus' *Contra Haereses*, but it is only firmly made in opposition to the Marcosians, who had contended that there was much to be known of the transcendent God from creation, apart from the revelation of God in Jesus.[72] For Bacon, who concerned himself directly with the things of nature, this distinction was a bit more crucial. As we have observed in the *Confession of Faith*, it was only through the mediator and his incarnation that true communication between the unknown God and creation was established. God could not be known personally through creation, apart from the act of uniting Himself with creation in the hypostatic union.

Regardless of the state of humanity in the Garden, Bacon's vision of recovery would be seriously impeded if the fall into sin meant that the human mind and soul had become so corrupt that they were prone to failure and wickedness. This point marks what is perhaps the most significant departure from Calvinism in Bacon's theology. For Calvin and his adherents, human knowledge still existed after the fall, but it was corrupt and always untrustworthy. For, as a part of the punishment for sin, "soundness of mind and uprightness of heart were withdrawn at the same time."[73] Although Calvin conceded

[69]   *Adversus Haereses*, Book 4, XX, 4; ANF, vol. 1, p. 488.

[70]   *Adversus Haereses*, Book 4, VI, 6; ANF, vol. 1, p. 469.

[71]   *Adversus Haereses*, Book 4, XX, 10; ANF vol. 1, p. 490.

[72]   *Adversus Haereses*, Book I, XVII; ANF, vol. 1, pp. 342–3.

[73]   ICR, Book 2, ch. 2, sec. 12. Translation from John Calvin, *Institutes of the Christian Religion*, trans. Ford Lewis Battles (Philadelphia, 1960), p. 270.

that all knowledge was not lost in the fall, and encourages thankfulness for the blessings which have, by God's "indulgence," proceeded from man's intellect in spite of its corruption, the real value of earthly knowledge is always suspect. Attempts to improve earthly conditions through arts and actions are ultimately vain, as is taught in the book of Ecclesiastes,[74] because corruption even adheres to the actions of God's chosen in this world.[75] Finally, the believer's proper attitude toward this earthly life itself must be one of renunciation, for heavenly life cannot be obtained if the soul is distracted by even the good things in a corrupt world.[76] As there can be no hope of transformation here, the hope of the believer must focus on transcending this world by passing to the next. The end result is that, while certain benefits may be derived from intellectual pursuits such as civil government and technological aids to life, the possibility of any actual recovery from the effects of the fall in this world is proscribed by the pervasiveness of the corruption it brought about. The Calvinist doctrine of total depravity applied to every aspect of human nature after the fall. As William Perkins put it, corruption of human nature was to be found "[i]n every part both of body and soul, like as a leprosie that runneth from the crown of the head to the sole of the foot."[77] Bacon differed sharply with Calvin and the Calvinists on this concept of the fall's net effect on human nature, and this, too, was significant for his understanding of the Instauration event. The key was the interpretation of Genesis 3:19, "In the sweat of thy face shalt thou eat bread."

Human Effort as the Key to Recovery

Lancelot Andrewes regarded Genesis 3:19 as a curse, but also as a verse that contained a blessing, and demonstrated the mercy of God: "God might have suffered the earth to have been fruitless let man have labored never so much, but that man for all his sinne, yet with his labour shall make the earth fruitful, in my opinion is a great mercy."[78] Andrewes considers it significant that God did not actually curse Adam at all in the fall, but cursed the earth instead.[79] As a result, "the fruitfulness must be recovered by man's labor, so that labor is a consequence of the earths Curse."[80] This is a point also made by Irenaeus in *Adversus Haereses*:

> For God is neither devoid of power nor of justice, who has afforded help to man, and restored him to His own liberty. It was for this reason, too, that immediately after Adam had transgressed, as the Scripture relates, He pronounced no curse against Adam personally, but against the ground, in reference to his works, as a certain person among the ancients has observed: 'God did indeed transfer the curse to the earth, that it might not remain in man.'[81]

---

[74]   Ibid., sec. 25, pp. 28–6.

[75]   Ibid., book 3, ch. 14, sec. 3. pp. 770–1. See also the mention of "Reason" being hopelessly blemished in William Perkins, *A Treatise of Man's Imaginations* (Cambridge, 1607), p. 149.

[76]   ICR, Book 3, ch. 9. Translation from Calvin, *Institutes*, trans. Battles, pp. 776–7.

[77]   William Perkins, *The Christian Doctrine*, edition in English and Irish (Dublin, 1652), p. 42.

[78]   Andrewes, Αποσπασματια *Sacra*, p. 320.

[79]   Ibid., p. 320: "but here the earth of which *Adam* was made, not *Adam* himself was cursed."

[80]   Ibid., p. 318.

[81]   *Adversus Haereses*, Book 3, ch. 23, 2–3; ANF, vol. 1, p. 456.

This concept has important implications for the potential of the intellect after the fall in Andrewes' theology, as opposed to the Calvinist doctrine of total depravity, and Bacon would take the idea a step further.

In the *Valerius Terminus* Bacon discussed the purpose of his program for the advancement of knowledge in terms of the recovery of the power which humanity possessed over nature in Eden:

> And therefore it is not the pleasure of curiosity, nor the quiet of resolution, nor the raising of the spirit, nor victory of wit, nor faculty of speech, nor lucre of profession, nor ambition of honour or fame, nor inablement for business, that are the true ends of knowledge; some of them being more worthy than other, though all inferior and degenerate: but it is a restitution and reinvesting (in great part) of man to the sovereignty and power (for whensoever he shall be able to call the creatures by their true names he shall again command them) which he had in his first state of creation. And to speak plainly and clearly, it is a discovery of all operations and possibilities of operations from immortality (if it were possible) to the meanest mechanical practice.[82]

In the same paragraph he discusses the precise extent to which this edenic power and knowledge may be recovered, and when it was recovered, in the age that he identified in the *Confession of Faith* as the third age of nature – namely the state after the fall:

> It is true, that in two points the curse is peremptory and not to be removed; the one that vanity must be the end in all human effects, eternity being resumed, though the revolutions and periods may be delayed. The other that the consent of the creature now being turned into reluctation this power cannot otherwise be exercised and administered but with labour, as well in inventing as in executing; yet nevertheless chiefly that labour and travel which is described by the sweat of the brows more than of the body; that is such travel as is joined with the working and discursion of the spirits in the brain ...[83]

The fall did not entail total depravity, nor did it result in a significant corruption of the human faculties as well as the human soul. The most significant change was that nature had rebelled against humanity, and now mastery could only be regained and subsequently maintained through great mental labor, which was the very purpose of the sciences. The other outcome of the fall was that, whatever humans accomplished, it would be rendered vain by the coming of the fourth age of nature, when there would be a new heaven and a new earth. Nevertheless, the "sovereignty and power" of Eden were recoverable. If Adam's ability to recognize the true name and purpose of the creatures on sight had also been lost, this was a setback. But it did not prevent man from obtaining a complete knowledge of all created things, for, as Bacon interpreted Ecclesiastes 3:11:

> Let no man presume to check the liberality of God's gifts, who as was said, *hath set the world in man's heart.* So was whatsoever is not God but parcel of the world, he hath fitted it to the comprehension of man's mind, if man will open and dilate the powers of his understanding as he may.[84]

---

[82] WFB, vol. III, p. 222.

[83] Ibid., pp. 222–3.

[84] Ibid., p. 221.

The human mind and its potential for knowledge remained as great as before the fall, and humans had the freedom to make the most of it if they so chose. This point is important for understanding Bacon's later statement in Aphorism 28 of the second book of the *Novum Organum* that human understanding is "depraved by custom and the common course of things," rather than by sin.[85]

Bacon's explanation of the Ecclesiastes 3:11 in *The Advancement of Learning* is slightly more extensive, and adds to our understanding of why the Instauration had not occurred before Bacon's own time:

> *God hath made all things beautiful, or decent, in the true return of their seasons: Also he hath placed the world in man's heart, yet cannot man find out the work which God worketh from the beginning to the end:* declaring not obscurely that God hath framed the mind of man as a mirror or glass capable of the image of the universal world, and joyful to receive the impression thereof, as the eye joyeth to receive light; and not only delighted in beholding the variety of things and vicissitude of times, but raised also to find out and discern the ordinances and decrees which throughout all those changes are infallibly observed. And although he doth insinuate that the supreme or summary law of nature, which he calleth *the work which God worketh from the beginning to the end*, is not possible to be found out by man; yet that doth not derogate from the capacity of the mind, but may be referred to the impediments, as of shortness of life, ill conjunction of labours, ill tradition of knowledge over from hand to hand, and many other inconveniences whereunto the condition of man is subject.[86]

Here again, the human mind is not the problem, but, after the fall, humanity suffers from shortness of life, and has, for one reason or another, failed not only to bring the labors of different people together, but also to establish a trustworthy tradition of human knowledge. There are other similar "inconveniences" and "impediments," but it is significant that, in the Instauration writings, Bacon gave directions for the removal of all these impediments; these ranged from the directions for removing the "Idols of the Mind" in the *Novum Organum* to his call for collective effort in the sciences to his suggestions for lengthening life indefinitely in the *Historia Vitae et Mortis*. Edenic mastery could be recovered, though, for various reasons, it had not been before.

This understanding of the fall remained constant throughout Bacon's writings. In the conclusion to the second book of the *Novum Organum* he wrote:

> For man by the fall fell at the same time from his state of innocency and from his dominion over creation. Both of these losses however can even in this life be in some part repaired; the former by religion and faith, the latter by arts and sciences. For creation was not by the curse made altogether and forever a rebel, but in virtue of that charter, "In the sweat of thy face shalt thou eat bread," it is now by various labours (not certainly by disputations or idle magical ceremonies, but by various labours) at length and in some measure subdued to the supplying of man with bread; that is, to the uses of human life.[87]

---

[85]　And therefore it can be mended by a change of method: "... *et medentur intellectui depravato a consuetudine et ab iis quae fiunt plerunque*," WFB, vol. I, p. 282.

[86]　WFB, vol. III, p. 265.

[87]　Spedding translation, WFB, vol. IV, pp. 247–8.

This twofold fall entailed a twofold solution: "Innocency" was restored by the action of the Church and faith, and dominion over creation was restored by the human work of Bacon's Instauration.[88] From this point onward in his sacred history the spiritual and material recoveries of humanity can be seen to proceed along separate, but thoroughly interrelated, paths: that of the incarnation and that of the Instauration.

Bacon's interpretation of Genesis 3:19, "In the sweat of thy brow shalt thou eat bread," figures prominently in both the *Valerius Terminus* and the *Novum Organum*, and it is important for understanding how he conceived of the Scriptural support for his argument. He has taken a verse which is commonly regarded as a curse, and has made of it not only a promise, but also a prophecy of human recovery. In doing so, he has established an interesting parallel between this verse and the combination of a curse and promise made to Eve. According to common Christian exegesis, childbirth would be painful and difficult for the woman as a result of the fall, but she had the assurance that eventually, through the bearing of children, the Messiah would come. The promise of the incarnation was made indirectly, as a result of the cursing of the serpent in Genesis 3:15 with the prophecy that his head would be crushed by the woman's seed. The prophecy of the Instauration was similarly indirect. Mastery of nature now required labor for the man, but eventually, in the Instauration, it would lead to recovery from the material loss brought by the fall. In the *De Augmentis*, Bacon made the charge that the "History of Prophecy" had been very much neglected by theologians. Genesis 3:19 appears to have been one verse which had not received its proper attention, and Bacon was willing to correct that omission.

Bacon went farther than Andrewes and Irenaeus in making Genesis 3:19 a prophecy of hope, but the difference was a matter of degree rather than substance. All was not yet known regarding the meaning of such prophecies as Genesis 3:19 or Daniel 12:4, but a new period was dawning, just before the second coming of Christ, when all such passages were becoming clear. When others came to understand what divine providence had been preparing for the world, the work of the Instauration would begin in earnest.

---

[88]   The reference to "idle magical ceremonies" in this passage may be referred specifically to the Paracelsians, for these are the ones who are censured by Bacon, in the second book of *The Advancement of Learning*, for assuming that the good effects may be brought about without labor, and to counter them Bacon employed his interpretation of Genesis 3:19. See his discussion in WFB, vol. III, p. 381.

Chapter 4

On the Way of Salvation:
Bacon's Twofold *Via Salutis*

From the day of the Genesis fall to the dawning of the seventeenth century, Bacon presented the tale of human history as the twofold recovery of what was lost in the fall. The narrative of sacred history leading up to the Instauration, as Bacon told it, included an explanation of why the Instauration had not occurred before his own age, and why the conditions were right for it to occur now. Bacon also had to account for why others before him had not recognized the divine plan for the Instauration. The answer to this is found in his call for a reform of the "History of Prophecy" and the "History of Providence." Before any of this could occur, Bacon had to abandon what had been one of the commonplaces of Western theology since the fourth century.

Bacon and Original Sin

Bacon's narrative of sacred history could not have been conceived if he had been bound to the view of sin that dominated the theology of Western Christianity in his day. By Western standards, Bacon's interpretation of history required a view of human nature that was heretically optimistic. The West, since Augustine, had embraced the idea of "original sin" as being an integral quality of human nature. Every human being, since the fall, had inherited this essential defect, with the exception of Jesus (and, for later Catholics, Mary). "Sin" was present in all people universally, generically, and necessarily. To varying degrees, in the West, sin tainted the entire human life, including intellectual activity, marking all things with an inherent corruption. If sin was an inherited and necessary handicap of the soul, then it followed that it was an insurmountable obstacle to recovery: like spiritual crabgrass, sin was universally present, corrupting all aspects of human nature, no matter what actions any individual might or might not have taken. For Calvin, recovery was precluded by the doctrine of total depravity in which man's intellect was corrupted in the fall, and no longer capable of correct, or uncorrupted, knowledge.[1] For Aquinas, and for most Western Christians who were not Calvinist, complete recovery was precluded not because the human reason itself was always corrupt, but because the ubiquitous sinful nature always derailed even the best efforts of the intellect.[2] Because human nature was inherently sinful, desiring evil things, it lacked the desire for virtue, which would make the works of the intellect

---

[1]   ICR, Book 2, ch. 12.
[2]   See Thomas Aquinas, *Summa Theologica*, vol. 26: Original Sin, 1a2ae, 81–85, ed. T.C. O'Brien (Cambridge, 2006), pp. 84–91.

pure in their effects. However, by Bacon's day, the Augustinian position of sin as an inherited and essential corruption was not without its detractors.

The Renaissance recovery of the Greek Fathers meant a sudden infusion of a very non-Augustinian concept of sin, which influenced both Lancelot Andrewes and Francis Bacon. For the Eastern Fathers, mankind was born weak and into a world of corruption. The weakness of the individual, and the tainted *environment* in which sin was ever present, meant that individuals, other than Jesus, would always commit sinful thoughts and acts, lacking the strength to do anything else. But sin was always regarded by the Christian East in its strict verbal sense: as the specific actions and thoughts of the individual who committed them, not as a quality of human nature itself.[3]

The interpretation of sin as the action of the individual, and not the quality of the human race, is precisely the position espoused by Lancelot Andrewes throughout his sermons.[4] For Andrewes, there was no generic or universal sinful nature which, *a priori*, limited knowledge for all. Rather, the actual sin of each individual was an obstacle to that individual's knowledge. Thus, when writing of the conditions which limited the knowledge of Solomon as an individual, Andrewes put it this way: "But these (mental qualities) were more excellently in Adam than in Salomon, who had no vanity to seduce him, no sicknesse to weaken him, no temptation to hinder his wisdome as Salomon had."[5] Similarly, in referring to Noah, Moses, and Solomon together, Andrewes wrote that "no one of them knew all things" as Adam did, again predicating the limitations of knowledge of each of them as individuals.[6]

In one sense, the implications for knowledge are not much different than those of Augustine's original sin doctrine, for all mortals who are born into a corrupt world are inherently too weak not to sin, and all will fail to recover edenic mastery through knowledge. However, there is also an interesting option left open when it is the shortcomings of each individual that limit knowledge: namely, the possibility that, through collective effort and correction, the errors of individuals could be overcome. Although this option was rejected for a number of reasons by Andrewes, as well as by the Eastern Fathers, Bacon adopted collaboration as an essential key to completing his program for the recovery of human knowledge.[7]

Andrewes' theology allows the possibility of a certain amount of recovery of edenic knowledge through method and art. This was achieved to a much greater degree by Noah, Moses, and Solomon than by the later heathen philosophers:

> The wisdome of all the Heathen Philosophers, compared to the knowledge of these three, *Noah, Moses,* and *Salomon,* was but ignorance. Yet *Adam* was created in wisdome, without corruption; their wisdome was bred in corruption, *and the Heathen are destroyed in their own Wisdoms, Psal. 9.15.* They three and all the wise men of the world had the light of their understanding *per scientiam acquisitam,* by study and former observation: *Adam* had

---

3   See Fr. John Meyendorff, *Byzantine Theology* (New York, 1974), pp. 143–6.

4   Nicholas Lossky, *Lancelot Andrewes, the Preacher (1555–1626): The Origins of the Mystical Theology of the Church of England,* trans. Andrew Louth (Oxford, 1991), pp. 168–76.

5   Lancelot Andrewes, Αποσπασματια *Sacra* (London, 1657), p. 214.

6   Ibid., p. 212.

7   See Bacon's discussion of this in the *De Augmentis,* WFB, vol. IV, p. 322 and also pp. 328–9. See also *Valerius Terminus,* WFB, vol. III, p. 231.

his without observation, *non per discursivam scientiam sed intuitivam*, for when he had beheld them he gave them names.[8]

We should note that, for Andrewes, it is not "wisdome" itself which is corrupt: even the knowledge of the godly men – Noah, Moses, and Solomon – came about in a corrupt world. It was a world in which nature was in a state of rebellion:

> We are here to note the obedience of the Creatures while man was obedient: and that the mutinie and discention between them, and their disobedience to man, did arise by mans rebellion to God his Maker.[9]

This concept of a rebellion in nature follows upon the rebellion of man and rests on an Aristotelian conception of a chain of causes in which humanity is the pivot point, or the *primum mobile*:

> Man is as the Great Sphear, the *primum mobile* to the other Creatures; his obedience to God draws the obedience of Plants, Trees, Beasts, and all the Elements unto him; during his obedience all Creatures are serviceable unto him; but afterwards the earth was unkinde, and as he moves all Creatures move with him: if he move against God all move against him.[10]

Knowledge is understandably greater among the godly patriarchs than among the heathen philosophers, according to this passage, for the patriarchs possessed the necessary prerequisite for regaining the obedience of created things – they were submissive and obedient to the one true God. Nevertheless, in a rebellious world, Noah, Moses, and Solomon had to come by their knowledge "*per scientiam acquisitam*, by study and former observation," while Adam had it from his creation. The difference between Andrewes and Bacon on the issue of recovery is not whether edenic knowledge *could* be recovered, but to what degree it *would* be recovered.

In the end, Andrewes focuses on man's spiritual recovery, rather than on the recovery of knowledge and power over creation. Where Bacon was optimistic concerning the potential of material recovery, Andrewes was pessimistic. Technological advances were the true vestiges of Adam's power, according to Andrewes, but these were not evidence of a new age, or even of improvement. For although, "the knowledge of the faith" in this time of Reformation, "is as the morning light which groweth lighter; the knowledge of reason is as the evening which groweth darker and darker."[11] Using the same basic theological assumptions, Bacon read the signs differently. The first evidence to assure Bacon that the Instauration of human mastery over nature was occurring was that of the biblical text, in which the patterns of God's saving actions were presented in prophecy.

---

8   Andrewes, *Αποσπασματια Sacra*, p. 212.
9   Ibid., p. 96.
10   Ibid., p. 318.
11   Ibid., p. 83.

Patterns in Divine Action and Prophecies of Instauration

Bacon's recognition of the coming of the Instauration event rested upon his firm belief in the consistency of God. Not only did God arrange nature according to laws which made it predictable, but in his providential actions he also proceeded according to consistent principles and patterns. Although God's actions were not predictable, except in so far as He revealed His intentions through prophecy, His consistency made the actions of the hand of providence recognizable to those who knew His ways. The central idea behind Bacon's plan for the "History of Providence," as he described the project in *De Augmentis*, was to document the observable actions of God and thus come to recognize the patterns of divine action. Thereafter, by continual observation, mankind might be more aware of the workings of God in the world.

Bacon had recognized the congruence of events which signaled the dawning of the providential age of the Instauration. One of the principles of providence was that the hand of God worked subtly, and that most people were entirely unaware of the significance of what was going on around them. As he had interpreted Luke 17:20: "And as it was said of spiritual things, 'The kingdom of God cometh not with observation,' so is it in all the greater works of Divine Providence; everything glides on smoothly and noiselessly, and the work is fairly going on before men are aware that it has begun."[12] When the pattern emerged, the full story of how God had intended, and prepared for, the Instauration all along could be clearly seen.

Bacon's understanding that God operates in patterns has significant ramifications for his use and interpretation of the Scriptures. The Protestant principle of biblical hermeneutics, which insists that the literal sense of Scripture is the essential one, and that it is singular, was not fully operational in the sixteenth and seventeenth centuries except among the Puritans.[13] It was certainly not observed by Bacon. Since God operated according to established patterns and principles in his providential interaction with the world, passages which applied to the incarnation could be readily applied to the Instauration. Luke 17 is just one example where the primary meaning of the text in its context referred to the incarnation, but could be extended to the Instauration because it set forth a basic theological principle. Matthew 22:29, "You err, not knowing the Scriptures nor the power of God," is another example.

According to the *De Augmentis Scientiarum*, Bacon saw it as erroneous to draw a one-to-one correspondence between prophecies and their specific fulfillment, for "though the height or fullness of them is commonly referred to some one age or particular period, yet they have at the same time certain gradations and processes of accomplishment through divers ages of the world."[14] It was an error, according

---

[12]   WFB, vol. IV, p. 92.

[13]   This is often assumed, in Protestant circles, to be the central difference between Protestants and Catholics in regard to biblical hermeneutics, but it was never so clearly articulated in the sixteenth and early seventeenth centuries, except in certain Puritan arguments. In the nineteenth century this position was erroneously read backward into Luther. For a concise summary of the issues involved see Steven Matthews, "Reading the Two Books with Francis Bacon: Interpreting God's Will and Power" in Peter J. Forshaw and Kevin Killeen (eds), *The Word and the World: Biblical Exegesis and Early Modern Science* (Basingstoke, 2007), pp. 61–77.

[14]   Translation, WFB, vol. IV, p. 313. Latin, WFB, vol. I, p. 515.

to Bacon, to assume that prophecies pertaining to human salvation were necessarily limited to the spiritual recovery accomplished through the incarnation. Bacon never interpreted the Scriptures as speaking solely of spiritual recovery. They were interpreted as the revelation of the will of God for the complete work of restoration, which comprehended the recovery of human sovereignty over nature as well. The narrative of the fall applied to the incarnation, but the same narrative applied directly to the Instauration. Similarly, prophecies pertaining to the restoration throughout the Old Testament were not to be rigorously limited to the incarnation or the parousia, as if they were not part of a greater package which was occurring continually in God's various acts of providence throughout time, all of which would only be ultimately fulfilled in the age yet to come. To his day, the significance of the incarnation had been well established, but the Scriptures had not yet been read for their full message of the restoration of fallen humanity. This was what Bacon called for in the reformation of the histories of providence and prophecy. And there were verses in the Scriptures which did apply directly to the Instauration.

We have already observed that, for Bacon. the fall was twofold, entailing both the loss of human "innocency" (or the state of grace) before God, and human sovereignty, or dominion over the creatures. Immediately after the fall, God issued promises, as well as curses. In the common Christian interpretation, Genesis 3:15 is the "protoevangelion," or "first gospel," as the words from God to the serpent concerning Eve's offspring are regarded as the first prophecy of the incarnation: "And I will put enmity between thee and the woman, And between thy seed and her seed; He shall bruise thy head, and thou shalt bruise his heel." As we have noted, Bacon also understood Genesis 3:19, "In the sweat of thy brow shalt thou eat bread," as a promise, and this was the first prophecy of the Instauration.

Other examples of direct Instauration prophecies would emerge as the "History of Prophecy" was developed, and as the divine pattern of the Instauration emerged. This method of interpretation followed a principle set forth by Irenaeus in the fourth book of *Adversus Haereses*:

> For every prophecy, before its fulfillment, is to men (full of) enigmas and ambiguities. But when the time has arrived, and the prediction has come to pass, then the prophecies have a clear and certain exposition.[15]

Even the prophecies of the coming of Christ in the incarnation were not properly understood before the event took place. This is evident from reading the Gospel of Matthew where the author goes to great lengths to show how the incarnation had been signaled in the Old Testament. It is noteworthy that Irenaeus made this statement in connection with verses from the Old Testament, which prophesied the increase of knowledge in the last days of the earth, and he included Bacon's favorite passage from Daniel. Bacon understood Daniel 12:4 to be exactly one of those prophecies discussed by Irenaeus, which would only be clear when it was coming to pass. In the *Valerius Terminus* Bacon explained that he felt "safe now after the event" to recognize the true meaning of Daniel 12:4, "Many shall go to and fro and

---

15   *Adversus Haereses*, Book IV, cap. 26, 1, ANF, vol. 1, p. 496.

science shall be increased," as applying to his own age.[16] With the coming of the Instauration, more of these dark prophecies would be made clear. Bacon found the Instauration clearly prophesied in the Psalms, as well.

One of the more misunderstood of Bacon's publications is his *Translation of Certaine Psalmes into English Verse* of 1625. This is partly because scholars have been too willing to interpret his words, in his dedication to George Herbert, that this work was "the poor exercise of my sickness" to mean that it was an idle pastime and had little to do with his more weighty work.[17] It is also a common mistake to assume that the Psalms in this collection are actually meant as translations out of another language. The language used for most of the versifications in this collection is entirely consistent with the Authorized, or "King James," version of the Scriptures, produced in 1611. In every case, when the texts of Bacon's translations and the Authorized version diverged, Bacon's text was not following any legitimate variant readings from another language – whether Greek, Hebrew, or Latin – but instead was interpreting the sense of the King James text, and adjusting the wording for rhyme and meter.[18] The Psalms were translated by Bacon into verse, not into English. This text is an act of biblical interpretation, in which the Psalms that Bacon has chosen are presented in light of the themes of the Instauration.

The 104th Psalm, replete with images of the interaction of God and the natural world, is the centerpiece of Bacon's *Translation of Certaine Psalmes*. In Jerome's edition of the Vulgate, as Charles Whitney has noted, verse 30 reads "*Emittes spiritu tuo et creabuntur; et instaurabis faciem terrae,*" although the more common reading of the Vulgate here has "*renovabis*" in place of "*instaurabis.*"[19] In the Authorized version this is rendered "Thou sendest forth thy spirit, they are created: and thou renewest the face of the earth." Bacon's interpretation of this verse was: "But when thy breath thou doth send forth again, / Then all things do renew and spring amain; / So that the earth lately desolate, / Doth now return unto the former state."[20] Bacon's interpretation presents the renewal of the earth as its restoration to a "former" fruitful state, as a result of a second emanation of God's creating breath, or his word. The sense of the recapitulation of a prior condition comes across clearly in Bacon's interpretation, though not necessarily in the original, and the image of a productive earth is foregrounded.

---

[16]   WFB, vol. III, pp. 220–21.

[17]   WFB, vol. VII, p. 274. Lisa Jardine and Alan Stewart give more than usual coverage of this work when they say, "In December 1624, Bacon published his *Apophthegms, New and Old*, and his *Translation of Certain Psalms*, in which he englished six or seven Psalms of David." (The correct number is seven.) See Jardine and Stewart, *Hostage to Fortune: The Troubled Life of Francis Bacon* (New York, 1999), p. 493.

[18]   In order to establish this I consulted the Septuagint, several versions of the Vulgate, and the critical edition of the Hebrew text in the *Biblia Hebraica Stuttgartensia* for each psalm. It was clear in all cases that Bacon was not using any of them as a primary text, but that his reading always followed the King James.

[19]   Charles Whitney, "Francis Bacon's Instauratio: Dominion of, and over, Humanity," *Journal of the History of Ideas*, 50/3 (1989), p. 377, fn. 14.

[20]   WFB, vol. VII, p. 283.

Another example of Bacon's reinterpretation of the Psalms is Psalm 90:13–17. A side-by-side comparison of Bacon's verses with the Authorized version is helpful for recognizing what Bacon has done:

Authorized version

13: Return, O LORD, how long? and let it repent thee concerning thy servants.
14: O satisfy us early with thy mercy; that we may rejoice and be glad all our days.
15: Make us glad according to the days wherein thou hast afflicted us, and the years wherein we have seen evil.
16: Let thy work appear unto thy servants, and thy glory unto their children.
17: And let the beauty of the LORD our God be upon us: and establish thou the work of our hands upon us; yea, the work of our hands establish thou it.

Bacon's interpretation[21]

Return unto us, Lord, and balance now,
With days of joy, our days of misery;
Help us right soon; our knees to thee we bow,
Depending wholly on thy clemency;
Then shall thy servants, both with heart and voice,
All the days of their life in thee rejoice.
Begin thy work, O Lord, in this our age,
Show it unto thy servants that now live;
But to our children raise it many a stage,
that all the world to thee may glory give.
Our handy work likewise, as fruitful tree
Let it, O Lord, blessed not blasted be.

Progress is added through the image of the blessed, fruitful tree, which, as we shall see elsewhere, was one of his standard metaphors for the Instauration. The original is a prayer for deliverance from hard times. Bacon's interpretation is that this is a prayer for the final relief of man's estate through progress in "our handy work." Far more than a mere idle pastime, Bacon's *Translation of Certaine Psalmes* presents the Instauration event as clearly prefigured in the prophetic words of the Psalms.

The Instauration in the History of Providence

As well as in the "History of Prophecy", evidence that the divine plan culminated in the Instauration was to be found in the "History of Providence," or the visible direction of history through the actions of God in the world. This too, according to Bacon, had not been adequately developed by the theologians up to his day. For Bacon, the hand of God had clearly been working to bring about the Instauration in the right season of

---

[21]   Ibid., p. 280.

the world, even as it had worked to produce the incarnation at the proper time. The key to understanding the history of providence as it pertained to the Instauration was recognizing why it had not occurred before, and why it could occur now.

In the *Valerius Terminus*, Bacon passed the following general judgment on the state of knowledge prior to the recorded learning of classical antiquity:

> For as for the uttermost antiquity which is like fame that muffles her head and tells tales, I cannot presume much of it; for I would not willingly imitate the manner of those that describe maps, which when they come to some far countries whereof they have no knowledge, set down how there be great wastes and deserts there: so I am not apt to affirm that they knew little, because what they knew is little known to us.[22]

This agnosticism concerning learning prior to recorded history is typical of Bacon's caution throughout his method when evidence was lacking, but it also reflects one of his most basic principles of chronology:

> ... for the truth is, that time seemeth to be of the nature of a river or stream, which carrieth down to us that which is light and blown up, and sinketh and drowneth that which is weighty and solid.[23]

If little had come down to his generation from "uttermost antiquity," much of the reason for it was that subsequent generations, such as the Greeks (who receive the blame for much of the loss of knowledge in Bacon's writings), were negligent in what they admired and consequently transmitted. One reason why knowledge had not increased in past ages is that these ages did not have their priorities in order, and they simply abandoned what they should have retained. There were other reasons, the most significant of which was that it was not yet time for the Instauration to occur. Other events had to take place first.

Bacon contended that the recovery of edenic knowledge had never occurred, and never could have occurred, prior to his own era, because knowledge, which he had described as "a plant of God's own planting,"[24] had not yet come into season:

> The encounters of the time have been nothing favourable and prosperous for the invention of knowledge; so as it is not only the daintiness of the seed to take, and the ill mixture and unliking of the ground to nourish or raise the plant, but the ill season also of the weather by which it hath been checked and blasted. Especially in that the seasons have been proper to bring up and set forward other more hasty and indifferent plants, whereby this of knowledge hath been starved and overgrown; for in the descent of times always

---

[22]   WFB, vol. III, p. 225.

[23]   *The Advancement of Learning*, WFB, vol. III, p. 292. See the repetition of this concern in the *Instauratio Magna*, where it is linked to the unfortunate democratic nature of knowledge, which served, in the past, to "dumb down" the legacy that was received: "*Quamobrem altiores contemplationes si forte usquam emicuerint, opinionem vulgarium ventis subinde agitatae sunt et extinctae. Adea ut Tempus, tanquam fluvius, levia et inflata ad nos devexerit, gravia et solida demerserit*" (WFB, vol. I, p. 127).

[24]   In *Valerius Terminus*, WFB, vol. III, p. 220.

there hath been somewhat else in reign and reputation, which hath generally diverted wits and labors from that employment.[25]

Bacon commonly used the metaphor of a fruitful plant to describe the providential development of an age in which knowledge would flourish, and to explain the reason why it had not done so before. Although the learning of those in the most ancient ages of the world might have been great, it had not produced the Instauration because it was not yet the proper season. Every age before Bacon's own had had other concerns, or hindrances, which had prevented the plant of knowledge from bearing fruit. In regard to antiquity, this led him to the judgment that there had been insufficient travel for the possibility of anything more than the beginning of the recovery of knowledge:

> But if you will judge of them by the last traces that remain to us, you will conclude, though not so scornfully as Aristotle doth, that saith our ancestors were extreme gross, as those that came newly from being moulded out of the clay or some earthly substance; yet reasonably and probably thus, that it was with them in matter of knowledge as the dawning or break of day. For at that time the world was altogether home-bred, every nation looked little beyond their own confines or territories, and the world had no through lights then, as it hath had since by commerce and navigation, whereby there could neither be that contribution of wits one to help another, nor that variety of particulars for the correcting of customary conceits.[26]

Antiquity was simply not the age in which "many shall go to and fro and knowledge shall be increased." At the earliest stages of human existence after the fall, the violence of the era reinforced this parochialism, and effectively prevented necessary travel:

> ... the studies of those times you shall find, besides wars, incursions, and rapines, which were then almost everywhere betwixt states adjoining (the use of leagues and confederacies being not then known), were to populate by multitude of wives and generation ... and to build sometimes for habitation towns and cities, sometimes for fame and memory monuments, pyramids, colosses, and the like.[27]

Bacon allowed that, in those early eras, there may have been much of use, which had simply been lost in the mud at the bottom of the river of time. The natural philosophy of past ages had not been preserved in the Scriptures any more than it had come down through an unbroken tradition, but the Scriptures still had much to say about the usefulness of, and the proper approach to, natural philosophy. Bacon found much of his history of providence in the biblical narratives.

From the first story after the fall, the account of Cain and Abel, Bacon drew support for his contention of the superiority of mental labor over brute labor, which we observed in his treatment of Genesis 3:19 in the *Valerius Terminus*. In *The Advancement of Learning* Bacon continued his sacred history after the fall as follows:

---

[25]   Ibid., pp. 224–5.
[26]   Ibid., p. 225.
[27]   Ibid.

> To pass on: in the first event or occurrence after the fall of man, we see (as the Scriptures
> have infinite mysteries, not violating at all the truth of the story or letter,) an image of the
> two estates, the contemplative state and the active state, figured in the two persons of Abel
> and Cain, and in the two simplest and most primitive trades of life; that of the shepherd,
> (who, by reason of his leisure, rest in a place, and living in view of heaven, is a lively
> image of the contemplative life,) and that of the husbandman: where we see again the
> favour and election of God went to the shepherd and not to the tiller of the ground.[28]

Abel was approved by God, not merely because of the nature of his offering, according
to Bacon's reading of Genesis 4, but also because of his devotion to contemplation
rather than brute labor. This is another example of Bacon's renovation of the "History of
Prophecy" where, as he claims, without contradicting the other truths which are taken
from these verses, he has added a prefiguring of the Instauration event, for it was not by
the sweat of the body, but by that of the brows, that mastery would be recovered.

Continuing his march through Genesis, Bacon noted next in *The Advancement of
Learning* that "in the age before the flood, the holy records within those few memorials
which are there entered and registered have vouchsafed to mention and honour the name
of the inventors and authors of music and works of metal."[29] Although the Scriptures
do not go into detail, according to Bacon it is clear from the fact that technological
advances were specifically mentioned in the Holy Scriptures that God approved of
such endeavors, particularly when the only other noteworthy feature of this epoch was
its violence and wickedness (which brought about the flood).

Following the biblical chronology, Bacon briefly mentioned the account of the
Tower of Babel in the next sentence:

> In the age after the flood, the first great judgment of God upon the ambition of man
> was the confusion of tongues; whereby the open trade and intercourse of learning was
> chiefly imbarred.[30]

Twice, on account of the wickedness of men, God had intervened in human advances,
and the recovery of power over nature was prevented. This is part of what Bacon
described as the "History of Providence," or, as he alternatively termed it, the
"History of the Judgments of God," in the *De Augmentis*. One of the hallmarks of the
true Instauration is that it coincides with the triumph of what Bacon regarded as "true
religion." It was the wickedness of the motive, after all, which was behind the fall
itself. Throughout his Instauration writings, Bacon emphasized the importance of the
proper motive – namely, charity, or, in practical application of that virtue, "the relief
of man's estate."[31] All other attempts at human achievement would be thwarted.

Next in line in Bacon's sacred history in *The Advancement of Learning* is "Moses
the lawgiver, and God's first pen."[32] Moses derived some of his excellence in natural
philosophy from the Egyptians, but there was a difference, which Bacon ascribed
to the observation of "some of the most learned Rabbins." According to the Jewish

---

[28]   Ibid., p. 297.

[29]   Ibid. See also *Valerius Terminus*, WFB, vol. III, p. 219.

[30]   WFB, vol. III, p. 297.

[31]   As per *The Advancement of Learning*, ibid., p. 294.

[32]   Ibid., p. 297–8.

authorities (and also Lancelot Andrewes, as we have noted above), Moses' knowledge was augmented by his piety, or devotion to the one true God. Therefore, Bacon contended (via vague references to Rabbinic authorities), the Law given through Moses reflected the wisest course according to both theology and natural philosophy. The case cited is the law concerning the isolation of lepers in Leviticus 13:

> As in the law of the leprosy, where it is said, *If the whiteness have overspread the flesh, the patient may pass abroad for clean; but if there be any whole flesh remaining, he is to be shut up for unclean*; one of them noteth a principle of nature, that putrefaction is more contagious before maturity than after: and another noteth a position of moral philosophy, that men abandoned to vice do not so much corrupt manners, as those that are half good and half evil. So in this and very many other places in the law, there is to be found, besides the theological sense, much aspersion of philosophy.[33]

The book of Job also demonstrated that, prior to classical antiquity there was significant concern for natural philosophy.[34] The example of Solomon is the pinnacle of pre-classical evidence for a proper and godly concern with natural philosophy. Having prayed for wisdom, God granted him knowledge of natural philosophy, as well as divinity and moral philosophy:

> By virtue of which grant or donative of God, Salomon became enabled not only to write those excellent parables or aphorisms concerning divine and moral philosophy, but also to compile a natural history of all verdure, from the cedar upon the mountain to the moss upon the wall ... and also of all things that breathe or move.[35]

Here, Bacon assumes that, just as there was a record of the proverbs of Solomon, there was once also a book of natural history written by Solomon, which was subsequently lost. It is a reasonable assumption given the parallels of wording in the verses of 1 Kings 4:29–33. In the examples that Bacon has chosen, there is a distinction between divine learning and natural philosophy, but the line of distinction is not one of absolute separation. Both Moses and Solomon were versed in divinity as well as natural philosophy, and Bacon has presented the two forms of knowledge as entirely complementary. As Bacon interpreted the case of Solomon, knowledge of both was the twofold result of his single prayer for wisdom, thus placing the origin of both in God. In the case of Moses, natural philosophy and theology inhere in the very same law, which may be understood according to either form of knowledge.

According to Bacon, knowledge of classical antiquity, as represented principally by the Greeks, was a poor affair when it came to natural philosophy. Although Bacon borrowed heavily from both Plato and Aristotle, and was not without praise for either, he described them as the flotsam and jetsam of time's river, compared to the weightier knowledge of earlier ages that had been lost.[36] Whilst Bacon's

---

[33] Ibid. We must note that Bacon is using an older definition of "aspersion" in which it connotes blessing rather than a negative.

[34] Ibid., p. 298.

[35] Ibid., pp. 298–9. See also the shorter parallel passage in *Valerius Terminus*, WFB, vol. III, pp. 219–20.

[36] As in Aphorism 77 of Book 1 of the *Novum Organum*, WFB, vol. I, p. 185.

objections to the methods of the Greeks, who, he claimed, favored disputation rather than experiment, have been frequently noted by Bacon scholars, there is also a decidedly religious reason for his criticism of the Greeks. Plato and Pythagoras, Bacon notes in different places, mixed their theology and their natural philosophy by using natural philosophy as an improper basis for their theology.[37] At the end of *Valerius Terminus* the "heathen" are censured for having a religion which consisted in "rites and forms of adoration, and not in confessions and beliefs." This led them to use natural philosophy as a platform for "metaphysical or theological discourse" and thus not to proceed further in the inquisition of nature itself.[38] In other words, the Greeks, and the heathen generally, did not have a functioning distinction between nature and the mysteries of God. This is stated succinctly in the *Novum Organum*, Book 1, Aphorism 79, where Bacon portrayed Greek natural philosophy as lasting a "minimum duration" [*minime diuturna*] before being eclipsed by "moral philosophy," which was as religion to the heathen.[39]

The Greeks have a significant place in the overall narrative of sacred history as a negative example. The Instauration could not have occurred among them because their erroneous religion contributed to their flawed approach to the method and meaning of natural philosophy. In keeping with his basic principle of associating proper human learning with what he regarded as the proper religion, Bacon lists many of his examples of the most learned of the pagan emperors of late antiquity as being amenable to, if not fascinated by, the Christian religion.[40] Yet, despite his negative comments, Bacon does note that the Greeks and Romans were aware of the divine nature of the work of natural philosophy and invention, and that this led them to revere their inventors and philosophers as gods.[41]

According to Christian theology, the incarnation of Christ is a singular event, but even this served to bolster Bacon's overall argument for the Instauration. The divinely ordained pattern of knowledge preceding action, which Bacon observed in the creation of light, and which underlay his argument for knowledge preceding action in the human activity in the Instauration, was also evident in the activity of Christ on earth. Bacon noted that, before working miracles, Jesus "did first shew his power to subdue ignorance, by his conference with the priests and doctors of the law."[42] This pattern of knowledge preceding works was also borne out in the coming of the Spirit at Pentecost: "And the coming of the Holy Spirit was chiefly figured and expressed in the similitude and gift of tongues, which are but *vehicula scientiae* [carriers of knowledge.]"[43] For Bacon, the incarnation was not the event which restored edenic mastery over nature, but one that reconciled humanity to God.

---

[37]   For Plato, see *The Advancement of Learning*, WFB, vol. III, p. 293, and the expanded version of this discussion in the *De Augmentis*, Book 2, WFB, vol. I, p. 570. Pythagoras serves as a somewhat more appropriate target in *Novum Organum*, Book 1, Aphorism 65, WFB, vol. I, p. 175.

[38]   WFB, vol. III, p. 251.

[39]   *Novum Organum*, WFB, vol. I, p. 187.

[40]   See *The Advancement of Learning*, Book 1, WFB, vol. III, pp. 305–7.

[41]   Ibid., p. 301. See also *Novum Organum*, Book 1, Aphorism 129, WFB, vol. I, p. 221.

[42]   WFB, vol. III, p. 299.

[43]   Ibid.

These examples are significant because they show that Bacon found his basic pattern of the proper order of creation inherent in the central act of the spiritual restoration of humanity. This pattern of knowledge preceding action was uncompromisingly paradigmatic. It was only natural that it should inhere in the Instauration as well.

Bacon observed that, after the Apostolic era, "many of the ancient bishops and fathers of the church were excellently read and studied in all the learning of the heathen."[44] He credited the Christian Church with preserving the knowledge of classical antiquity through the barbarian invasions, but did not ascribe to the Church Fathers any advancement of natural philosophy. There was a very good reason for this, as Bacon explained in the *Novum Organum*:

> Now it is well known that after the Christian religion was received and grew strong, by far the greater number of the best wits applied themselves to theology; that to this both the highest rewards were offered, and helps of all kinds most abundantly supplied; and that this devotion to theology chiefly occupied the third portion or epoch of time among us Europeans of the West; and the more so because about the same time both literature began to flourish and religious controversies to spring up.[45]

Concerns for the faith came first. This was the era in which doctrine was sorted out and codified, and the Christological controversies were settled by the ecumenical councils. Ten aphorisms later, Bacon noted that a few Church Fathers were hostile to certain basic conclusions of natural philosophy concerning the roundness of the earth and the antipodes.[46] This, however, derived only from their religious zeal, and, according to Bacon, served to demonstrate that, even in the proper religion, misguided zeal can be a hindrance to natural philosophy.[47] It was a rhetorical reminder to the zealous divines in his own society that hostility to the conclusions of natural philosophy was unnecessary, for none in Bacon's day was bothered by a round earth or the consequent existence of antipodes.

Moving closer to his own era, Bacon regarded the medieval Scholastics as having a decidedly negative effect on both theology and natural philosophy. Having taken the erroneous Aristotle as their "dictator" in the sciences, they could not profit from extensive travel or a wide variety of written sources, "as their persons were shut up in the cells of monasteries and colleges." As a result, their learning dissolved into "a number of subtle, idle, unwholesome, and (as I may term them) vermiculate

---

[44] Ibid., p. 299.

[45] WFB, vol. I, p. 187; Spedding translation, WFB, vol. IV, p. 78. By the third epoch of time Bacon means after the era of pagan Rome, which was the second era. This section must be carefully observed within the context of Aphorism 79. The aphorism itself censures those who have willfully neglected natural philosophy, but this part is not a censure, but an exemption of the Christians. Bacon resumes the censure with the philosophers of the second age in the next lines.

[46] WFB, vol. I, p. 196, Aphorism 89.

[47] In his notes to the Latin version Robert Leslie Ellis drew attention to the fact that this section is strikingly similar to a passage in Kepler's *De Stella Martis* in which the Fathers are cited as Lactantius and Augustine. It seems reasonable that Bacon may have been simply acknowledging a commonplace concerning the ignorance of certain Church Fathers rather than establishing a point of his own.

questions, which have indeed a kind of quickness and life of spirit, but no soundness of matter or goodness of quality."[48] In the *Novum Organum* the Scholastics are passed over, along with the "Arabians:"

> For neither the Arabians nor the Schoolmen need be mentioned; who in the intermediate times rather crushed the sciences with a multitude of treatises, than increased their weight.[49]

The problem with the Scholastics is nothing other than the sin of pride, which led them to depart from both of God's two books – the book of scripture and the book of nature:

> But as in the inquiry of the divine truth their pride inclined to leave the oracle of God's word and to vanish into the mixture of their own inventions, so in the inquisition of nature they ever left the oracle of God's works and adored the deceiving and deformed images which the unequal mirror of their own minds or a few received authors or principles did represent unto them.[50]

For the reading of both books, a time of reform was at hand.

Martin Luther was the pivotal figure for the reformation of the reading of Scripture, even as Bacon, to his own way of thinking, would restore the proper reading of nature. To accomplish his task, Luther had to turn to the ancient authorities, as there was little sympathy for his position in his own time:

> Martin Luther, conducted (no doubt) by an higher Providence, but in discourse of reason finding what a province he had undertaken against the Bishop of Rome and the degenerate traditions of the church, and finding his own solitude, being no ways aided by the opinions of his own time, was enforced to awake all antiquity, and to call former times to his succors to make a party against the present time; so that the ancient authors, both in divinity and in humanity, which had long time slept in libraries, began generally to be read and revolved.[51]

Although it is true that Luther turned to ancient authorities to counter the arguments of his own age, it is interesting that Bacon has laid the Renaissance phenomenon of the recovery of the texts of classical antiquity, and not just the Church Fathers, solely at the feet of Martin Luther and his Reformation interests. In the continental translation of this passage, Tobie Matthew, by then a practicing Roman Catholic, modified it considerably, and made Luther a *part* of the phenomenon rather than its cause.[52]

According to Bacon's narrative, Luther's appeal to ancient texts instituted the important scholarly trend of a careful reading of texts, involving meticulous

---

[48] *The Advancement of Learning*, WFB, vol. III, p. 285. The use of "vermiculate," as well as the image of liveliness amidst "no soundness of matter," refers to Bacon's earlier comparison of this learning to "many substances in nature which are solid [and] do putrefy and corrupt into worms." See also the censure in the same on p. 288.

[49] *Novum Organum*, Book 1, Aphorism 78, WFB, vol. I, p. 186; Spedding translation, WFB, vol. III, p. 77.

[50] WFB, vol. III, p. 287.

[51] WFB, vol. III, pp. 282–3.

[52] See Spedding's footnote at ibid., p. 283.

attention to grammatical detail in the original languages. However, this concern for accurate reading had soon produced a negative result, "for men began to hunt more after words than matter," and even that learning which the Scholastics did possess "came to be utterly despised as barbarous."[53] From Bacon's perspective this concern for words over substance in the Reformation was just a recent example of a long-standing "distemper of learning" which had plagued mankind and prevented progress throughout past ages. Now, in Bacon's own age, this was changing, and all things were being set right, starting with the observable changes in religion. However, this reformation was accompanied by another:

> And we see before our eyes, that in the age of ourselves and our fathers, when it pleased God to call the church of Rome to account for their degenerate manners and ceremonies, and sundry doctrines obnoxious and framed to uphold the same abuses; at one and the same time it was ordained by the Divine Providence that there should attend withal a renovation and new spring of all other knowledges: and on the other side we see the Jesuits, who partly in themselves and partly by the emulation and provocation of their example, have much quickened and strengthened the state of learning.[54]

Human learning, along with divine learning, was already undergoing a transformation according to the providence of God. It could be observed by those sensitive to the course of the history of providence. It could be seen in the Jesuits, whom God, according to his providence, had used as a spur to the work of others, as well as agents of change in their own right. Bacon linked the reform of the Church with the reform of human learning, both of which were occurring gradually, "in the age of ourselves and our fathers," but in 1605 they were both underway. Bacon felt that he stood on the threshold of a new era, as was evident by the obvious workings of the hand of providence, and supported by the proper reading of many passages of Scripture. The history of prophecy and the history of providence, so often neglected, both pointed toward a providential age in which humankind would finally recover mastery over nature.

In Bacon's own age the proper conditions had arisen, by God's providence, for the investigation of nature and the advancement of technology, and the human race stood poised on the verge of a new era of the "Kingdom of Man," [*Regno Hominis*] as he termed it in the title of the second book of the *Novum Organum*. The benefits of the discoveries and technological advances which came from his program would "extend to the whole race of man,"[55] and would result in the restoration and extension of the "dominion of the human race over the universe."[56] The whole program for the restitution of the sciences, while it was to be the project and product of human agency, would nonetheless take place in accordance with the divine plan and will, and would be governed by proper religion:

---

53 Ibid., pp. 283–4.
54 Ibid., p. 300.
55 *Novum Organum*, Book 1, Aphorism 129, WFB, vol. I, p. 221.
56 Ibid., p. 222.

> Only let the human race recover that right over nature which belongs to it by divine bequest, and let power be given it; the exercise thereof will be governed by right reason and sound religion.[57]

The Instauration was a matter of divine and human co-operation in which, quite understandably, a "sound" or "healthy" religion, was an essential element. Other elements were the opening of the world to travel, and relative civil peace and prosperity. According to Bacon, his was the age predicted by the prophet Daniel.

Bacon's Providential Age

Daniel 12:4, *multi pertransibunt et augebitur scientia*, in the Vulgate, or "many shall pass to and fro, and science shall be increased," in Bacon's own translation, is the most prominent Scriptural support for Bacon's belief in the recovery of human knowledge. On the lavish frontispiece of the 1620 *Instauratio Magna* it serves as a caption for the image of tall ships passing back and forth through the pillars of Hercules, signifying that the old barriers and limitations were no longer in place. There is a specific section in each of the major Instauration writings – *Valerius Terminus*, *The Advancement of Learning*, the *Instauratio Magna*, and the *De Augmentis* – in which Bacon explains the meaning of this verse. The earliest example of his exegesis, and the true prototype for his later treatments, is the following passage from the *Valerius Terminus*:

> This is a thing which I cannot tell whether I may so plainly speak as truly conceive, that as all knowledge appeareth to be a plant of God's own planting, so it may seem the spreading and flourishing or at least the bearing and fructifying of this plant, by a providence of God, nay not only by a general providence, but by a special prophecy, was appointed to this autumn of the world: for to my understanding it is not violent to the letter, and safe now after the event, so to interpret that place in the prophecy of Daniel where speaking of the latter times it is said, *Many shall pass to and fro, and science shall be increased;* as if the opening of the world by navigation and commerce and the further discovery of knowledge should meet in one time or age.[58]

"Now after the event," Bacon wrote, it is safe to recognize that this apocalyptic passage was a special prophecy referring to the Instauration. The Instauration was already underway, and Bacon's generation was in the middle of it. In this "autumn of the world," knowledge, planted and tended by God through the ages, was bearing fruit. With the event at hand it was also safe to look back over the whole of sacred history, and recognize the threads of this divine plan laid throughout the Scriptures and evident in the history of God's providential action through the ages. If others had not recognized the traces and evidence of the coming of this new era as he had, that

---

[57]   Ibid., p. 223: "*Recuperet modo genus humanem jus suum in naturam quod ei ex dotatione divina competit, et detur ei copia: usum vero recta ratio et sana religio gubernabit.*" I have adjusted the Spedding translation to remove the confusion of adjectives in the final clause. Cf. WFB, vol. IV, p. 115.

[58]   WFB, vol. III, pp. 220–21.

was not particularly surprising, for as he explained in the parallel discussion of this verse in the *Novum Organum*:

> Now in divine operations even the smallest of beginnings lead of a certainty to their end. And as it was said of spiritual things, "The kingdom of God cometh not with observation," so is it in all the greater works of Divine Providence; everything glides on smoothly and noiselessly, and the work is fairly going on before men are aware that it has begun.[59]

As we observed earlier, the text on which Bacon grounds the principle of the unobtrusiveness of providential activity is Luke 17:20. Most of his audience would have been aware that the words are Christ's, and refer not just to "spiritual things," but also specifically to his own coming in the flesh. The Pharisees had asked when the Kingdom of God would come, not having seen the signs, and had recognized that the King himself was right before them. As in the incarnation, so also in another of God's major acts – the Instauration – the full truth of the prophecies was not recognized until they were being fulfilled.

There is more to Luke 17:20. It is a transitional verse which not only applies to the first coming of the Christ, but also opens an extensive discussion of the nature and conditions of the parousia, or the second coming of Christ at the culmination of the present, and final, age of the world. Significantly, this verse does not foretell the last moments of the earth as marked by gloom and despair, as some other passages have been interpreted (Matthew 24, for example), but instead emphasizes the coming of Christ as a complete surprise. Up until the very end people would be going about their business and daily lives, without suspecting that their labor and activity would soon be cut off. Suddenly, and as unexpectedly as the lightning (Luke 17:24), Christ would return and put an end to all, and the new heaven and the new earth would ensue. The presentation of the eschaton here is entirely supportive of the providential age which Bacon believed he was observing, both in the prophetic words of the Old Testament and in the obvious course of human events. The providential age of the Instauration was marked by continual human labor and effort, until all human labor and accomplishments would be ended, for "vanity must be the end in all human effects, eternity being resumed."[60] However, there is nothing in this passage which would proscribe the reduction, or even elimination, of the suffering and misery produced by material causes. If the parousia is so surprising, it must, at the very least, mean that the human condition is not particularly desperate. The Instauration event signified an upturn in the human condition in the autumn of the world. It was to be a true "*valerius terminus*" – a strong or healthy ending.

According to a key hermeneutical principle of the Reformation – that scripture interprets scripture, meaning that the clear or obvious passages shed light on the more difficult passages – Bacon is on solid ground here, and he could make a valid claim to consistency according to the standards of his society. For it was manifest, according to Bacon, that such an age of material recovery was at hand, and hence prophecies which might be taken to suggest otherwise would therefore be in need of

---

[59]   WFB, vol. I, p. 200; Spedding translation, WFB, vol. IV, pp. 91–2. See the parallel use of this verse, in its original form, in *Valerius Terminus*, WFB, vol. III, p. 223.

[60]   WFB, vol. III, p. 222.

re-examination. The clear evidence of his world pointed to the end time being as he read it in Luke 17 and Daniel 12:4. Bacon saw the convergence of a constellation of essential conditions for the Instauration all around him.

The Conditions for Instauration

According to the *Valerius Terminus*, one condition for the Instauration was the "opening of the world by navigation," and this had occurred in the past two centuries with the voyages of discovery and the opening of new trade routes. What Bacon said of the limitations of "uttermost antiquity" was not true of his own age: that it was "home-bred, and every nation looked little beyond their own confines or territories ... there could neither be that contribution of wits one to help another, nor that variety of particulars for the correcting of customary conceits."[61] In his day, that which had prevented a proper "conjunction of labours,"[62] was no longer an issue: one of the impediments to recovery had been removed. In addition, the voyages of discovery had provided a more complete knowledge of the world itself. They might be joined with the labors of the continental Jesuits as part of the increase in knowledge which was already underway.

Another condition for the Instauration was civil peace and prosperity. Before classical antiquity, all ages also had to contend with the violence of their times, for they were marked by "wars, incursions, and rapines, which were then almost every where betwixt states adjoining (the use of leagues and confederacies being not then known.)"[63] This meant that human knowledge was occupied with concerns of survival, rather than advancing itself and dominating nature. Bacon, as he saw it, was living in a time and place which was notably free from such disruptions. Under Elizabeth, England had experienced "constant peace and security," and under James the "felicity in the people" was connected with "learning in the prince."[64] Although Bacon's constant praise for King James was certainly part of the patronage discourse, it is no less true that he saw the continual peace under Elizabeth and James, as well as his own considerable advancement under James, as part of the providential arrangement for the Instauration. It is not mere flattery but an aspect of his understanding of the history of providence when, in the dedicatory epistle of the 1620 *Instauratio Magna*, Bacon stated that his work was "as a child of time rather than of wit," and that "if there be any good in what I have to offer, it may be ascribed to the infinite mercy and goodness of God, and to the felicity of your Majesty's times."[65]

---

[61]   Ibid., p. 225.

[62]   *The Advancement of Learning*, ibid., p. 265. From the *De Augmentis* we may note that it is this conjunction of labors which "supplies the frailty of man (*mortalium fragilitati succurrit*)" (WFB, vol. I, p. 486).

[63]   From *Valerius Terminus*, WFB, vol. III, p. 225.

[64]   *The Advancement of Learning*, Book 1, WFB, vol. III, p. 307.

[65]   Latin, WFB, vol. I, p. 123: "*pro partu temporis quam ingenii*" and "*ut si quid in his quae affero sit boni, id immensae misericordiae et bonitati divinae et foelicitati temporum tuorum tribuatur*," respectively. Translation, WFB, vol. IV, p. 11.

As the Instauration was an event which required the preparations of divine providence, Bacon always held that it was not to be credited to his own ingenuity. This is not to suggest that he saw himself as playing anything less than a pivotal role in the advent of this providential age. It was he who had observed the hand of God in the course of events and recognized that the fulfillment of Daniel 12:4 was at hand. And it was he who, as a result of the "infinite goodness and mercy of God," had struck upon the proper method for the complete knowledge of nature. According to the *De Augmentis Scientiarum*, the most important element for the advancement of the sciences was prudent and sound direction (*consilii prudentia et sanitas*): "For the cripple in the right way (as the saying is) outstrips the runner in the wrong."[66] Bacon was the mere instrument of providence for providing humanity with the proper guidance for the recovery of dominion over creation. He was but the means to a divine end. Thus he wrote in the preface to the *Instauratio Magna*:

> Wherefore, seeing that these things do not depend upon myself, at the outset of the work I must humbly and fervently pray to God the Father, God the Son (Word), and God the Holy Ghost, that remembering the sorrows of mankind and the pilgrimage of this our life wherein we wear out days few and evil, they will vouchsafe through my hands to endow the human family with new mercies.[67]

Clearly, Bacon saw himself as a part of God's preparation for the new era. In the preface to the *Novum Organum* he described his object as "being to open a new way to the understanding, a way by them untried and unknown." The adherence to "parties and schools," which had marked learning in previous eras, was "at an end," and Bacon was "merely as a guide to point out the road; an office of small authority, and depending more upon a certain kind of fortune than upon any ability or excellency."[68] If any should wonder why previous generations, steeped in their errors, had not seen this new road of human learning before, Bacon responded in Aphorism 78 of the *Novum Organum* that it was more surprising that anyone should have seen the right way, given past errors. But, again, this was to be attributed not to Bacon's intellect, but to things occurring in due season – a point which is clear in the Latin but not the English:

---

[66]   *De Augmentis Scientiarum*, Book 2, WFB, vol. I, p. 486. Translation, WFB, vol. IV, p. 284.

[67]   WFB, vol. I, p. 131. Translation, WFB, vol. IV, p. 20.

[68]   The Latin of the passage in its entirety reads: "*Verum quum per nos illud agatur , ut alia omnino via intellectui aperiatur illis intentata et incognita, commutata jam ratio est; cessant studium et partes; nosque indicis tantummodo personam sustinemus, quod mediocris certe est authoritatis, et fortunae cujusdam potius quam facultatis et excellentiae*" (WFB, vol. I, p. 153; Spedding translation, WFB, vol. IV, p. 41). I have corrected the translation in my own rendering, most notably in translating "*fortunae cujusdam*" as "a certain kind of fortune," rather than "a kind of luck." The sense of extreme arbitrariness implied by "luck" is not inherent in "*fortunae cujusdam*," which, given the qualification of "*cujusdam*" does not mean "sheer luck" or "accident." Given the ubiquity of the ascription of the Instauration to divine providence everywhere except where he is referring to his own role, this appears as simply a case of appropriate lack of presumption. Writing "God chose me for this" would be sure to draw fire.

... ut tollatur omnis admiratio, haec quae adducimus homines hucusque latuisse et fugisse;
et maneat tantum admiratio, illa nunc tandem alicui mortalium in mentem venire potuisse,
aut cogitationem cujuspiam subiisse: quod etiam (ut nos existimamus) felicitatis magis
est cujusdam, quam excellentis alicujus facultatis; ut potius pro temporis partu haberi
debeat, quam pro partu ingenii. [... that all wonder how these considerations which I
bring forward should have escaped men's notice till now, may cease; and the only wonder
be, how now at last they should have entered into any man's head and become the subject
of his thoughts; which truly I myself esteem as the result of some happy accident, rather
than of any excellence or faculty in me; a birth of Time rather than a birth of Wit.][69]

Spedding's translation has been left unaltered, but his choice of "some happy accident"
for '*felicitatis ... Cujusdam*' is unfortunate. The sense is not so "accidental" in the
Latin. The word *felicitatis* can be taken as "good fortune," but it must be borne in mind
that it relates to *felix*, which means "fruitful" or "fertile." *Felix*, like *felicitatis*, can also
mean "fortunate," or "successful," but both words always imply the fortune which
results from proper conditions, and due season. An apple tree full of apples in the fall
can be regarded as a "fortunate" or "successful" occurrence (since apple trees might
not bear fruit), but it is hardly an accident or a matter of mere chance. In addition, the
last phrase of this passage ('*ut potius pro temporis partu haberi debeat, quam pro*
partu ingenii') is clearly epexegetical, explaining (particularly in light of the "birth"
imagery) that Bacon's realizations occurred because the time was right for them.

It is significant, in the course of Bacon's sacred history, that the Instauration
followed the Reformation, for, according to Book 1 of the *Novum Organum*, "sound
religion," as well as proper reason, was required for the governance of that power
over nature which was mankind's divine bequest.[70] More than fifteen years earlier,
in the *Valerius Terminus*, Bacon had presented the priority of religion over human
knowledge as one of the foundational principles of his program, mandating, "[t]hat all
knowledge is to be limited by religion, and to be referred to use and action."[71] Sound
religion was another element in the constellation of conditions for the Instauration,
which divine providence had arranged through the Reformation.

It is also significant that Bacon did not regard the Reformation as something
fully accomplished by Luther or Calvin, but as a process being carried out "in the
age of ourselves and our fathers." The concept of an ongoing Reformation is often
associated with the Puritan call for a continuing purge of papal abuses. Neither
Bacon nor Andrewes believed that the Reformation was complete for, as Andrewes
said, "the knowledge of the faith is as the morning light which groweth lighter."[72]
But this light came from the East. It is noteworthy that in *The Advancement of*
Learning Bacon identified the significance of Luther's role in the Reformation with
the recovery of antiquity. As the Eastern fathers were recovered, and the faith of the
early Christians became better known, proper religion was being restored by people
such as Bacon and Andrewes.

---

[69]   WFB, vol. I, p. 186; Spedding translation, WFB, vol. IV, p. 77.

[70]   See above, as well as the originals of this passage, in WFB, vol. I, p. 223; Spedding
translation, WFB, vol. IV, p. 115.

[71]   WFB, vol. III, p. 218.

[72]   Andrewes, *Αποσπασματια Sacra*, p. 83.

Bacon did not leave the course of the Reformation solely in the hands of the professional theologians. In calling for the establishment of the disciplines of the history of prophecy and the history of providence, Bacon assumed an active role in the process of theological reform. We have already observed the significance of these new disciplines for Bacon's understanding of the Instauration as an event in sacred history. But "sound religion" also required the establishment of the proper line between the investigation of nature, for which the human mind had been designed, and the *Deus Absconditus*. According to Bacon, past confusion on this point had been harmful to both natural philosophy and religion, and this had been one of the impediments to the Instauration in former generations.

In his draft of a discussion of "the impediments which have been in the state of heathen religion and other superstitions and errors of religion," near the end of the *Valerius Terminus*, Bacon explained that Christianity had the "singular advantage" over other religions of recognizing the boundary line of the *Deus Absconditus*:

> And of the singular advantage which the Christian religion hath towards the furtherance of true knowledge, in that it excludeth and interdicteth human reason, whether by interpretation or anticipation, from examining or discussing of the mysteries and principles of faith.[73]

The heathen religions, as we noted in regard to Bacon's opinions of the Greeks, and Plato and Pythagoras in particular, had erred by founding religion on natural philosophy. In this same section Bacon also derided the "abuse of Christianity" which led theologians to limit or proscribe natural philosophy for fear of prying into the mysteries of God.

This paragraph from the *Valerius Terminus* is a recapitulation of a lengthy discussion earlier in the same work on God, religion, and the study of nature. Bacon's statement of the principle "[t]hat all knowledge is to be limited by religion, and to be referred to use and action"[74] led off this earlier discussion, in which he separated the knowledge of the "will" and "nature" of God, which cannot be learned or observed in "sensible and material things," from the study of creation, which in its entirety was intended by God to be subject to human investigation. There is no distinction here between sacred and secular studies, for Bacon firmly maintained throughout this discussion and his Instauration corpus that the study of creation served to reveal a great deal about its Creator. But the study of creation did not reveal everything. Creation revealed God's power, glory, and principles of order, as opposed to His transcendent nature (the "divine essence" of Patristic theology), and His will. The distinction here is a theological one, which finds repeated expression throughout the Instauration corpus in Bacon's interpretation of Matthew 22:29:

> For, saith our Saviour, *You err, not knowing the Scriptures nor the power of God;* laying before us two books or volumes to study if we will be secured from error; first the Scriptures revealing the will of God, and then the creatures expressing his power.[75]

---

73   WFB, vol. III, p. 251.
74   Ibid., p. 218.
75   Ibid., p. 221. See also the parallel passages in *The Advancement of Learning*, WFB, vol. III, p. 301, and the *Novum Organum*, Book 1, Aphorism 89, WFB, vol. I, p. 197.

Regarding the "two books" it is important to consider the timing. Through the Reformation the proper reading of the first book was being restored. With "sound religion" in place as a guide, the time to open the second book was at hand.

Sound religion was also a necessary prerequisite for the Instauration because a proper attitude, or disposition of the heart, was required of those working toward the understanding of nature, if their work was not to be derailed by human vanity. Throughout the Instauration writings, Bacon emphasized the need for the Christian virtues of charity and humility to guide the efforts of mankind in the Instauration. Knowledge, for Bacon, did not have power as its end and goal, as Hobbes claimed. Rather, knowledge *was* power – a power given to man by God – and it was to be wielded for the ends which God intended. As Bacon wrote in the *Valerius Terminus*:

> But yet evermore it must be remembered that the least part of knowledge passed to man by this so large a charter from God must be subject to that use for which God hath granted it; which is the benefit and relief of the state and society of man; for otherwise all manner of knowledge becometh malign and serpentine, and therefore as carrying the quality of the serpent's sting and malice it maketh the mind of man to swell; as the Scripture saith excellently, *knowledge bloweth up, but charity buildeth up*. And again the same author doth notably disavow both power and knowledge such as is not dedicated to goodness or love, for saith he, *If I have all faith so as I could remove mountains* (There is power active,) *if I render my body to the fire,* (There is power passive,) *if I speak with the tongues of men and angels,* (There is knowledge, for language is but the conveyance of knowledge,) *all were nothing.*[76]

Here Bacon has taken the serpent imagery, used by his Calvinist opponents to denounce knowledge, which "puffeth up," and has explained it according to St Paul's argument for the importance of charity in 1Corinthians 13. It was not knowledge that was the problem, according to Paul, but the absence of charity as the proper motive and end of knowledge. Without this motive in place, knowledge *would* be malign and result in human pride, which was the cause of man's fall in the first place. Charity and humility were closely connected. Bacon's chaplain, William Rawley, in a sermon dedicated to Bacon, placed them both under the heading of "Meekness" in connection with the Sermon on the Mount reference, "blessed are the meek."[77] Even so, for Bacon, charity and humility were essential to the Great Instauration, through which man, even in the age of hardship after the fall, would "inherit the earth."

If charity and humility were absent it would be impossible for the patient labor necessary for human recovery to occur:

> The access to this work hath been by that port or passage, which the divine Majesty (who is unchangeable in his ways) doth infallibly continue and observe; that is the felicity wherewith he hath blessed an humility of mind, such as rather laboureth to spell and so by

---

[76]   WFB, vol. III, pp. 221–2. See also the discussion of the value of man imitating the goodness of God in the same work at pp. 217–18. For comparison, see the passages on charity in *The Advancement of Learning*, WFB, vol. III, p. 266; also Book 2, p. 421; and the *Instauratio Magna*, WFB, vol. I, pp. 131–2. This understanding of charity is consistent throughout, though it is presented in greatest detail, and with the most explicit Scriptural exegesis, in the *Valerius Terminus*.

[77]   William Rawley, *A Sermon of Meeknesse* (London, 1623).

degrees to read in the volumes of his creatures, than to solicit and urge and as it were to invoke a man's own spirit to divine and give oracles unto him. For as in the inquiry of divine truth, the pride of man hath ever inclined to leave the oracles of God's word and to vanish in the mixture of their own inventions; so in the self-same manner, in the inquisition of nature they have ever left the oracles of God's works, and adored the deceiving and deformed imagery which the unequal mirrors of their own minds have represented unto them.[78]

Again, the theme of the two books undergirds Bacon's argument, and he equates the heretical error of misreading the Scriptures with the misinterpretation of the "volumes of the creatures," even as he did in the *Meditationes Sacrae*. Both are the result of pride and human interests eclipsing divine truth.

Throughout the *Novum Organum* Bacon presented the flaws in human thinking and perception which had prevented recovery in previous eras as "idols of the mind." This is usually regarded in scholarly works as merely a convenient metaphor to express the idea that past errors are the result of human fantasy, or that they are particularly difficult to dislodge because they have been enshrined and regarded as incontrovertible through long use. Whilst there is no doubt that Bacon intended both of these meanings to be conveyed, he never suggested that the choice of the term "idols" was merely metaphoric. In the passage from *Valerius Terminus* just quoted, errors in natural philosophy were nothing less than misreading one of God's two books, and adoring the false images of nature constructed in the human mind. There is something genuinely idolatrous at work in the adherence of past generations to the idols of the mind. Later, in the sixteenth chapter of *Valerius Terminus,* Bacon discussed the concept of idols more directly, and associated the various idols of the mind with the idolatrous error of predicating a human form of the eternal and transcendent God. The errors in both cases were the result of "pride and partiality as well as of custom and familiarity," for humans were projecting their own understandings and mental images on that which was outside them, rather than learning about things as they really were.[79] Idolatry was an error of perception. The human mind, created to be a "glass" that reflected and comprehended the cosmos, was instead forcing God's creation into its own preconceptions, even as idolaters recast God in their own image. These cherished preconceptions were the "idols of the mind." Without the humility to accept things as God had made them, and the selfless desire to turn knowledge to the holy end of charity, pride and idolatry would follow.

Francis Bacon regarded himself as standing at the beginning of the "Autumn of the World." This was to be the age of fruitfulness when the plant of God's own planting, knowledge, came into season and its benefits were harvested by human hands for the relief of the suffering which had marked human existence since the fall. Human beings were to be the stewards of God's bounty, as they were in the Garden. Human mastery over creation would be restored in the time of the "Kingdom of Man" which would precede the final coming of the Kingdom of God. A divine symmetry was being accomplished in time: In the beginning, the Word had descended from his eternal existence to create the cosmos and place humanity in the Garden to rule over it, but this was interrupted by the fall and an age of suffering. Now, near the end of

---

[78]   WFB, vol. III, pp. 223–4.
[79]   WFB, vol. III, pp. 241–2.

time, human suffering would wane, the conditions of the Garden would be restored, and then, at the close of the age, eternity would resume. For Bacon, this basic pattern was established by the God who had created a universe of order in the first place. But the actual harvest was still ahead, and the course which the Instauration would take from here would depend on human agency and conscious choice as much as on the divine will. As God once ruled creation through human beings as His viceregents, so the beneficial order of creation would be restored through humans in co-operation with His will.

Chapter 5

In the Autumn of the World:
Features of the Age of Instauration

By his own estimation, Francis Bacon stood on the verge of the age of material recovery predicted in the prophecies. He was close enough in time to the event that he could clearly see many features of the Instauration that had been obscured to previous generations. But he was not without guides for his interpretation. His description of the new era, in terms of what it would mean and what it would involve, had a precedent and a pattern: In reading the prophecies, Irenaeus of Lyons had seen many of the same things; and as God always worked according to patterns, the earthly order of the Instauration had a model in the order of the Church.

Irenaeus and Francis Bacon on the Golden Age

Irenaeus of Lyons believed that, at the end of the age in which he lived – the age of suffering – humanity would be restored to its rightful place of lordship over the natural world. In other words, creation would be restored to its state before the fall – the state of fruitfulness in service to humanity – and the meek would receive their inheritance.. He based this conclusion on the same verse that William Rawley, Bacon's chaplain, used as the central text in his only published sermon: "Blessed are the meek, for they shall inherit the earth" (Matt. 5:5). We will return to Rawley's sermon in the next chapter, as it reflects the influence of Bacon's theology of Instauration. For now, we must consider where that theology originated. According to Irenaeus, this was made clear in Paul's letter to the Romans, which stated:

> For the expectation of the creature waiteth for the manifestation of the sons of God. For the creature has been subjected to vanity, not willingly, but by reason of him who hath subjected the same in hope; since the creature itself shall also be delivered from the bondage of corruption into the glorious liberty of the sons of God.[1]

The delivery of the creatures would occur in conjunction with a time in which Christ would fulfill the prophecy of David in Psalm 104:30, "and renew the face of the earth."[2] Life in this renewed earth would be marked by a material prosperity in which

---

[1]   *Adversus Haereses*, Book V, cap. 32, ANF, vol. 1, pp. 561–2. The translation of Romans 8:19–21 is taken from the ANF, representing how Irenaeus understood these verses, rather than the Authorized version translators.

[2]   *Adversus Haereses* Book V, cap. 33, 1, ANF, vol. 1, p. 563.

grapes and grain would produce many thousandfold more than in ages before,[3] and in which the animals would be returned to their edenic subjection to man.[4]

As in so many other places, it is difficult not to recognize the parallels between Bacon and Irenaeus concerning this last golden age of the earth. Throughout his writings, Bacon rested his theological points on many of the same scriptural passages used by Irenaeus to make the same points, and the connection of Psalm 104:3 to the Instauration is another telling example. Yet there was also a significant difference between the two visions of recovery.

In Bacon's day, Irenaeus was not only popular but also controversial, partly because of a recent development in the publishing of his texts. As Charles Whitney has observed, the last five chapters of Irenaeus' *Adversus Haereses* had only very recently been recovered in Western Europe.[5] Prior to the 1575 edition of Francis Fuardentius the sections dealing with the restoration of nature and human governance over the earth were missing from Western editions, and for an understandable reason.[6] These chapters revealed that Irenaeus' golden age was to be a genuine millennium: an age when Christ would come and reign on earth for a thousand years, and in which the saints of all ages who had died would be raised in the "first resurrection" to rest from their labors and reign with Him over creation.[7] In Irenaeus' time such a literal millennialism was, in the words of Jaroslav Pelikan, "a mark neither of orthodoxy nor of heresy," but "one opinion among others within the range of permissible opinions."[8] Very early in the medieval period, such a literal reading of the apocalyptic passages had fallen out of favor with mainstream Christian thinkers and had been replaced by the more allegorical reading of a "thousand years" espoused by Augustine.[9] Irenaeus' work appears to have been edited accordingly.

In Bacon's England the concept of a literal thousand-year reign of the saints after the resurrection was associated with the most radical and puritan strains of

---

[3]   This passage in Irenaeus is a more likely source for Bacon's grape and grain imagery in the Feast of the Family in the *New Atlantis* than the pagan fertility cults to which David Innes assumed that it must refer. See David C. Innes, "Bacon's New Atlantis: The Christian Hope and the Modern Hope," *Interpretation*, 22/1 (1994), p. 22. The grapes and grain also relate to the Eucharistic promise that the feast of Holy Thursday would be resumed in the earthly Kingdom of Christ (as Irenaeus interpreted it). This was when the accomplishments of the faithful would be celebrated. See *Adversus Haereses*, Book V, cap. 33, 1. In this, too there is a relation to the Feast of the Family as Bacon presented it.

[4]   *Adversus Haereses*, Book V, cap. 33, 3–4, ANF, vol. 1, p. 563.

[5]   Charles Whitney, *Francis Bacon and Modernity* (New Haven, 1986), p. 44.

[6]   On the absence of the final chapters and the editions of Fuardenti see *Patrologia Graeca*, vol. 7, pp. 10–11.

[7]   *Adversus Haereses*, Book V, cap. 32–35. There has been some debate about how literally Irenaeus should be read in regard to the period of a thousand years precisely, but this debate did not exist in Bacon's era. In light of *Adversus Haereses*, Book, V, cap. 28, 3, the literal gap of a thousand years after the "first resurrection" seems the most likely reading.

[8]   Jaroslav Pelikan, *The Christian Tradition* (5 vols, Chicago, 1971), vol. 1, p. 125.

[9]   See ibid., vol. 3, pp. 42–3. Interpretations of the Council of Ephesus which claim that Chiliasm was clearly condemned there in 431 are highly debatable. At best, *certain types* of millenarian thinking were condemned. Others did quite well long afterward.

Reformation thought. Bacon's Instauration, in contrast with the golden age of Irenaeus, lacked any of the marks by which it could be identified with the genuinely millennial thought of his day. There is no sense anywhere in Bacon's writings that he shared Irenaeus' idea that the restoration of human mastery over nature would be part of a thousand-year period, and Bacon certainly did not regard this as happening after a "first resurrection" of the righteous. For him, the Instauration was something that was taking place in an imperfect world right in front of him. The divinely appointed age of Instauration could legitimately be called an "apocalyptic" age, but in the ancient Greek sense of an "apocalypsis" as a process of unveiling or revealing. In the Instauration event an important, but previously obscure, aspect of the divine plan was becoming manifest. Dark meanings that had been hidden in the Scriptures had come to light, and the many twists of history which had once seemed insignificant could now be recognized as the accomplishment of so many "divine counsels" leading toward this age. Bacon took Irenaeus' theology and adjusted it towards a more orthodox perspective. This adjustment could be ascribed to the view of the gradual revelation of prophecy which Bacon shared with Irenaeus. Irenaeus was writing about the fulfillment of prophecies in the distant future, whereas Bacon believed that he was currently witnessing that fulfillment, and, as Irenaeus himself had acknowledged, the meaning of prophecy was enigmatic and ambiguous until that time when it actually came to pass. Rather than pointing toward a true millennium, Bacon's understanding of the prophecies is in keeping with the concept of "inaugurated eschatology" which dominated Eastern and early medieval Western theology after the decline of Irenaeus' more literal apocalypticism.

Inaugurated Eschatology in Bacon's Instauration

Augustine and his early medieval followers shared with the Greek Fathers a basic understanding of the nature of time after the incarnation. Georges Florovsky has labeled this concept of time "inaugurated eschatology."[10] This idea has very recently become popular among theologians in the West, largely as a result of Florovsky's influence, but it was not common in Western Europe in the sixteenth and seventeenth centuries. It was noticeably *present* by virtue of the recovery of the Eastern Fathers and because of a certain lingering tradition in the West, but during the later medieval period it had been largely eclipsed by the more rational explanations of the Scholastics. Nevertheless, it was a key feature of Lancelot Andrewes' theology, and it is yet another aspect of Bacon's system which he appears to have shared with Andrewes.

Inaugurated eschatology is commonly described as the tension between "even now" and "not yet" in the life of the Church during the time between the resurrection and the second coming of Christ. The "end times" were inaugurated by Christ when he was present in the flesh after the incarnation, but the fullness of the benefits of heaven are not yet realized until the second coming. Behind this doctrine is the concept that in the Church, through participation with Christ (who as God transcends time),

---

[10]   Georges Florovsky, *The Collected Works of Georges Florovsky* (14 vols, Belmont, MA, 1987), vol. 1, p. 36. This is the essay in which Florovsky introduces the term, though he expands it throughout his writings.

past, present, and future are to some degree transcended, and the normal barriers of time become permeable. Because of the uniting of time-bound humankind with the timeless God through the mediation of Christ, even in this life some of the benefits and blessings of the future life are realized, though imperfectly. Thus, the Eucharist is not a "memorial" in the sense that it is merely a time for reflection on a past event, and neither is it, as Calvin claimed, merely a pledge of the feast to come.[11] It is the feast itself. It is exactly the same feast that Christ ate with his disciples on Holy Thursday, and it is exactly the same feast that is underway eternally in heaven. It is all one experience, although the fullness of the experience, in which the feast is celebrated free from the taint of sin, and in the visible presence of the Savior, only exists beyond time, in the "age to come."[12] A typical example of this doctrine is found in the sermon of the fourth-century Father, John Chrysostom, which is read annually at the feast of Pascha in Eastern churches to this day. Chrysostom invites his audience to the eternal feast, taking place at that very moment in the liturgy, in time-transcendent language: "Enter ye all, therefore, into the joy of our Lord, and let both the first and those who come after partake of the reward."[13] The implications of this Patristic view of time were by no means limited to the Eucharist: the Church, though not yet perfect, is even now the Kingdom of God; human beings, though not yet sinless, even now are "freed from sin" and have been restored to their proper status before God; and even now something of the eternal peace and glory of God which pervades heaven can be experienced by the believer, though the experience is imperfect.

Nicholas Lossky has demonstrated that the language of inaugurated eschatology is present throughout Lancelot Andrewes' sermons. This can clearly be seen in the following passage from a sermon on the resurrection based on Colossians 3, which he takes quite literally:

> It is an error certainly, which runneth in men's heads when they hear of the Resurrection, to conceive of it as of a matter merely future, and not to take place till the latter day. Not only "Christ is risen", but if all be as it should be, "We are already risen with Him", saith the Apostle, in the Epistle this day, the very first words of it; and even here now, saith St. John, is there a "first resurrection", and happy is he that "hath his part in it".[14]

The words from St John's Apocalypse (20:5–6), are a source text for those millennialists who, like Irenaeus, believed that there would be a "first resurrection" in which

---

[11]   ICR, Book 14, ch. 14, sec. 8, and 12.

[12]   The Eastern Patristic concepts in regard to the sacraments are elaborated more fully by Florovsky in his collection of essays on Creation and Redemption, (Florovsky, *Collected Works*, vol. 3). See in particular the essays on "Redemption" and "The 'Immortality' of the Soul." The distinctions between this Eastern and the more common Western views are clarified somewhat in his final essay in this volume, "Eschatology."

[13]   See this section of the sermon in the theological context of the idea that Christians enter into the "eighth" or timeless, "day of creation" in Vladimir Lossky, *The Mystical Theology of the Eastern Church* (Crestwood, NY, 1976), pp. 247–9.

[14]   Lancelot Andrewes, *Works of Lancelot Andrewes*, ed. John Parkinson (11 vols, Oxford, 1854), vol. 2, p. 199. The verse is also used by Nicholas Lossky for this point: see Nicholas Lossky, *Lancelot Andrewes, the Preacher (1555–1626): The Origins of the Mystical Theology of the Church of England*, trans. Andrew Louth (Oxford, 1991), pp. 164–5.

the saints would be raised to rule on earth with Christ, before the second, general resurrection on the last day. Unlike the millennialists, Andrewes understood this first resurrection in terms of inaugurated eschatology: it was taking place already in the Church, and the apocalyptic age of this first resurrection began with the Church at Pentecost. Even at that time, those dead in sin had been raised into life in the Kingdom which is the Church. They were not yet free from all pain and suffering, though they were experiencing a foretaste of the full joy of heaven by virtue of the sacraments. Bacon's Instauration was also possible in the "here and now," in an imperfect way.

One aspect of Bacon's Patristic shift, as we observed in Chapter 2, was his modification of the language of time and eternity. Another important aspect was his understanding that the incarnation occurred primarily for the purpose of uniting God and creation mystically through the mediator, Christ, and that, in the Church, Christ was already participating with creation through His flock. In this way, inaugurated eschatology was declared to be part of his personal creed in his *Confession of Faith*. In a passage that we have already seen from *Valerius Terminus*, Bacon presented the limitations of the Instauration in terms of the essential tension of "even now" and "not yet:"

> It is true, that in two points the curse is peremptory and not to be removed; the one that vanity must be the end in all human effects, eternity being resumed, though the revolutions and periods may be delayed. The other that the consent of the creature now being turned into reluctation, this power cannot otherwise be exercised and administered but with labour, as well in inventing as in executing …[15]

Even now, humanity could regain mastery over nature, though it would not yet be possible without continual labor. Even now, humankind could be relieved from the physical suffering which had dominated human existence since the fall, but this relief was not yet perfect and permanent. The benefits of the new heaven and the new earth were present, though not fully realized. In the Church the restoration of the spiritual "innocency" was already being accomplished. After the advent of the Instauration, dominion over creatures was also taking place. This understanding can profitably inform our reading of Bacon's conclusion to the *Novum Organum*, where he stated:

> For man by the fall fell at the same time from his state of innocency and from his dominion over creation. Both of these losses however can even in this life be in some part repaired; the former by religion and faith, the latter by arts and sciences.[16]

Laborers in the Fields of Instauration: Orders and Offices

If the Kingdom of Man could only be established through labor, it was necessary to have laborers. In one sense, this charge was given to all of humanity in Bacon's writings. Throughout his writings pertaining to the reform of learning Bacon subsumes the human race under the typically biblical heading of "man," and it is clear that all are to reap the rewards of the Instauration. But even as the Church, where, according to 1 Peter 2:5, all members were in the "priesthood," had need

---

[15]   WFB, vol. III, pp. 222–3.

[16]   Spedding translation, WFB, vol. IV, pp. 247–8.

of a special class of clergy to tend it, so also the Instauration was to be largely the work of those set aside for the task.[17] Bacon understood experimentation as a holy vocation, and as no less than a work of a special holy order. In the second book of *The Advancement of Learning* he argued that there must necessarily be a fraternity of those dedicated to learning, as a consequence of its holy source:

> And surely as nature createth brotherhood in families, and arts mechanical contract brotherhoods in communalities, and the anointment of God superinduceth a brotherhood in kings and bishops; so in like manner there cannot but be a fraternity in learning and illumination, relating to that paternity which is attributed to God, who is called the Father of illuminations or lights.[18]

Although the specific goal of this section is to argue for greater co-operation across political boundaries for the advancement of the sciences, the justification which Bacon uses presents those engaged in the sciences as having a divine mandate for their activity. This was God's work, and those who did it did the work of God no less than kings and bishops who held their offices from God.

Bacon understood his own vocation as the advocate for the Instauration in terms of a religious office: "And thus I conceive that I perform the office of a true priest of the sense (from which all knowledge in nature must be sought, unless men mean to go mad) and a not unskilful interpreter of its oracles."[19] Although he shied away from using the term *sacerdos* in the original of this passage, preferring the basically synonymous *antistites religiosos*, his pious young friend George Herbert did not avoid the term. In a poem written in Bacon's honor Herbert described him as "the singular priest of the world and of souls" [*mundiquae & Animarum, sacerdos unicus*].[20] Both uses may appear to be mere metaphor were it not for the general context of the Instauration as a divine work and the further evidence of the *New Atlantis*.

As Francis Bacon's only work of fiction, *New Atlantis*, has been the subject of countless literary analyses exploring the meaning of the imagery. These studies very properly analyze the scientific activities of "Salomon's House" on the island of Bensalem as a model for what Bacon expects to see in the Instauration. Unfortunately, far too many scholars have also made a further assumption that does not stand the test of historical context and textual evidence: namely that Bacon was anything but a sincere Christian, and that the centrality of scientific activity in the *New Atlantis* is a bid for elevating secular activity over religious observance. The *New Atlantis* was published posthumously by the respected theologian Dr William Rawley, Bacon's former chaplain and amanuensis. As Bacon's friend and literary collaborator, the

---

[17] The "Priesthood of All Believers" is often mistakenly assumed to be a distinctly Protestant doctrine. Although extreme forms of it certainly marked the more radical elements of the Reformation, as it is found in the Lutheran Confessions, this doctrine is one on which the Lutherans and the Catholics basically agreed. At Augsburg the question between the two sides was not one of whether all were in a sense "priests," nor of having a professional order of clergy, but the status of ordination.

[18] WFB, vol. III, p. 327.

[19] WFB, vol. I, pp. 138–9; translation, WFB, vol. IV, p. 26.

[20] George Herbert, *The Works of George Herbert* (New York, 1880), p. 587.

Reverend Dr Rawley understood Bacon's work better than many later critics, and knew that an attack on Christianity was not part of it. Rawley described Bacon's intent in this work in his introduction to the published version:

> This fable my Lord devised, to the end that he might exhibit therein a model or description of a college instituted for the interpreting of nature and the producing of great and marvellous works for the benefit of men, under the name of Salomon's House, or the College of the Six Days' Works.[21]

We can restore the proper reading of the *New Atlantis* if we recognize that the allegories and metaphors it contains reflect Bacon's theology, which he stated more clearly in his non-fiction works.

In light of the theological system which permeates Bacon's Instauration writings, it is clear that the scientific activity of Salomon's House was a religious observance, just as similar activity had been for Adam in the Garden. This makes sense out of the sacerdotal images associated with the order of men inhabiting Salomon's House. Bacon describes those in Salomon's House an "Order" dedicated to "the study of the Works and Creatures of God."[22] As the reader comes to learn more of this order, it becomes apparent that it is structured on the pattern of an episcopal hierarchy with the "Father of Salomon's House" acting in the place of the *episcopos* – literally the "overseer" – of the divine activity of the order. When this dignitary first makes his appearance he is clothed in the manner of a rather well-off Dominican priest with the white undergarment and black outer garment typical of the Order of Preachers. The symbols of his office, the same as those borne by bishops – a crosier and a pastoral staff – are carried before him.[23] Those of the order who are seen by the travelers to Bensalem are always similarly vested. Just as Bacon saw it as necessary within the Church to have some, like monastics, who were dedicated to a life of prayer, the Order of Salomon's House is dedicated to a life of studying God's works. There are even those who have adopted the life of hermits in order to read the book of nature.[24] From top to bottom, the structure of Salomon's House is presented as a typical religious order complete with "novices and apprentices," and an outer circle of servants analogous to a confraternity, or possibly the village of lay servants and associates which often surrounded the monastery.[25] Life within the order is structured by regular prayer, both of thanksgiving, and "imploring his aid and blessing for the illumination of our labours, and the turning of them into good and holy uses."[26] Although the work was left unfinished, the last image we have of the Father of Salomon's House is of him laying his hands on one kneeling before him, and giving his blessing.

Just how much of the ceremonial and hierarchical order of Salomon's House Bacon thought would, or should, be put into practice is questionable – *New Atlantis* is, after all, an imaginative work of fiction. But from the imagery we can get a strong

---

[21]   WFB, vol. III, p. 127.
[22]   Ibid., p. 145.
[23]   Ibid., pp. 154–5.
[24]   Ibid., p. 157.
[25]   Ibid., p. 165.
[26]   Ibid., p. 166.

impression of the sense of divine vocation of those in the "fraternity in learning and illumination" described in *The Advancement of Learning*.

Rebuilding the Temple of Nature

Throughout Bacon's philosophical writings he describes the Instauration activity in terms that recall the rebuilding of Solomon's temple by the Children of Israel after the Babylonian exile. As the true holy place of worship was once restored, so also the true holy knowledge was now being restored, along with the vocation that accompanied this knowledge. These themes have been well examined by Charles Whitney, and more recently Stephen McKnight.[27] In light of the present discussion the most significant point is that the temple imagery reinforces the principle that God always operates according to patterns, and that the description of one recovery could be readily applied to another.

Both Whitney and McKnight identify a particular parallel between Solomon's temple and the Instauration that is worthy of consideration here, since it fits very well with the overarching incarnation/Instauration parallelism which governs Bacon's theology. This parallel is based on the observation that the Solomon's temple, as part of its function as the chief place of sacrifice and the ritual center of the world for the people of Israel, was designed to be a model of the whole cosmos, or, quite literally, a "microcosm." Whitney has expressed this point most succinctly:

> Temples are almost universal symbols for the world, *imagines mundi*, as Mircea Eliade says, and the iconography and architecture of churches and cathedrals shows. Solomon's Temple is supposed to have contained the pattern of the universe within it.[28]

It is also common to run across explanations of the images of flora and fauna in churches and in Solomon's temple, which describe the place of worship as a figurative restoration, or model, of the Garden of Eden. The two interpretations are non-contradictory, as Eden was commonly seen as a microcosm, representing the whole cosmos as the ritual center of creation where God walked with humanity. From a historical point of view, it is difficult to ascertain just how clearly places of worship, or specifically Solomon's temple, were seen in this recapitulatory manner at any given time in the intellectual history of Europe. This theme is ubiquitous in the architecture of churches and cathedrals, and occasionally it is clearly expounded in literature. One extensive discussion of this theme, noted by Whitney, is the study of the Jerusalem temple published by the Jesuits, Hieronymo Prado and Juan Bautista Villalpando in their widely-read *In Ezechiel explanationes et apparatus vrbis, ac templi Hierosolymitani*, which appeared between 1596 and 1605.[29] Spanning three massive and lavishly illustrated volumes it is an in-depth discussion of the temple as microcosm.

---

[27]   Whitney, *Francis Bacon and Modernity*; Stephen McKnight, *The Religious Foundations of Francis Bacon's Thought* (Columbia, 2006).

[28]   Whitney, *Francis Bacon and Modernity*, p. 33.

[29]   (Rome, 1596–1605).

We cannot be sure that Bacon read the work of Prado and Villalpando, although after 1607 he had ready access to both continental philosophy and the Jesuit order through the now exiled Father Tobie Matthew. There were, of course, other possible sources as well. The real problem is that we only have one rather enigmatic passage from Bacon that directly suggests that he would have embraced the interpretation of the temple as a model of the cosmos. In Aphorism 120 of the first book of the *Novum Organum* Bacon presented his intentions for his Instauration writings in terms of constructing a "holy temple" rather than a secular or pagan monument: "And for myself, I am not raising a capitol or pyramid to the pride of man, but laying a foundation in the human understanding for a holy temple after the model of the world."[30] Without the understanding of the temple as *imago mundi* this is a very difficult passage to interpret indeed.

The view of the temple as a model of the world meshes completely with Bacon's understanding of the Logos as the mediator between God and all of creation. Solomon's temple was the dwelling place of the special presence of God. Although the entire universe could not contain the Creator, according to Solomon's speech in 1 Kings 8:27, he had promised to meet his chosen people in the special place of the temple for the rituals which united them to himself.[31] If the temple is also the model of the cosmos, then this is the place where the mediator would come to figuratively unite with the cosmos, and foreshadow the permanent unity with the cosmos that would occur in the incarnation. There is a resonance here with Irenaeus' statement that Solomon built the temple as "the type of truth," meaning that it prefigured the truth which would be revealed in the incarnation of the Logos.[32]

According to the *Novum Organum* quotation, the location for the temple which Bacon was building was the human intellect (*intellectu humano*). This should not be taken to suggest that the temple itself is imaginary – merely internalized. If the Solomon's temple is recognized as a model of the cosmos, we have another significant parallel between incarnation and Instauration: in both cases the temple undergoes a process of internalization.

Solomon's temple was the place of sacrifice for the people of Israel. According to Christian teaching from the first century onward, the physical temple was supplanted by the body of Christ, as stated in John 2:19–21, and the sacrifices carried out at the temple were rendered obsolete by the singular sacrifice of Christ on the cross, according to Hebrews 9:11–12. The "temple," from the incarnation onward in Christian theology, is not to be found in Jerusalem, but wherever Christ is to be found, including in the believers themselves, as Paul declared in 1 Corinthians 3:16. In this way, the physical temple has been exchanged for a spiritual temple in regard to the propitiatory or salvific function. It has been internalized. There would still be a need for special times and places of worship in the early Church, but not specifically for Solomon's temple. In much the same way, if the temple was a physical image of the world when it stood in Jerusalem, Bacon is proposing that this function of the temple be internalized

---

[30]   WFB, vol. I, p. 214; translation, WFB, vol. IV, pp. 106–7.
[31]   Leviticus 16:2.
[32]   *Adversus Haereses*, Book 4, XXVII, 1, ANF, vol. 1, p. 499.

as well. The human intellect is now to be the location where the true image of the world – of which the old temple was a prefiguration – may be found.

Human Agency and the Instauration

The Instauration, as Bacon conceived it, was both a divine action and the product of human effort. Human agency and free will were also essential features of the work of Instauration. Guided by charity, humanity joined its Maker in an event which would accomplish the divine end of mercy, relieving the suffering of all humankind. In his *Confession of Faith* Bacon stated, "God created man in his own image, in a reasonable soul, in innocency, in free-will, and in sovereignty."[33] According to Bacon's sacred history, human beings lost precisely two of these in the fall: innocency, and sovereignty. They retained reasonable souls and free will, both of which were essential for the part they were to play in the Instauration. "Man" must, according to "sound reason," "open and dilate his powers as he may" to recover his sovereignty over the created order. Humans were to recover their original mastery, exercised by Adam in Eden. If this was a genuine sovereignty over lesser things, it would require genuine human agency, not the agency of a puppet whose actions were controlled by the immediate providence of God.

From beginning to end, the Instauration writings present human beings as the agents of their own recovery, if they would but choose to set out upon the new way which God had prepared, and which Bacon was illuminating. This had not occurred before Bacon's day because the human will had not been bent toward the proper ends, according to Aphorism 97 of the first book of the *Novum Organum*: "No one has yet been found so firm of mind and purpose as resolutely to compel himself to sweep away all theories and common notions, and to apply the understanding, thus made fair and even, to a fresh examination of particulars."[34] God had now removed the obstacles which had previously derailed human efforts, but humanity must still have the fortitude to turn away from all previous notions and assumptions about the cosmos, and, by observation of the natural world, relearn the basic principles of nature. This required human labor – the sweat of the brow. If God had opened the doors of the Instauration, it was still for humankind to step through. In the autumn of the world humans would come to a correct understanding of the role of their own free choice in effecting recovery.

Human choice and agency were not only essential for explaining the Instauration in the context of Bacon's sacred history, they were also necessary for the method itself. The investigation of nature proceeded according to the choices made by the investigators, who decided what to investigate and determined what conclusions should be drawn. According to Bacon's interpretation of Solomon's words in Proverbs 25:2, the role of God was to conceal a thing, but the role of a "king," or human being, who was created for sovereignty, was to seek it out. Therefore it is not at odds with the idea of the Instauration as an act of divine providence for Bacon to take it upon

---

[33]   WFB, vol. VII, p. 221.

[34]   WFB, vol. IV, p. 93. Latin, WFB, vol. I, p. 201. See also Aphorism 94, WFB, vol. I, p. 200; translation, WFB, vol. IV, p. 92.

himself to construct the method by which his fellow laborers in the Instauration could come to understand the divinely established laws of nature. He was performing his proper role as the human instrument by which the Instauration would come about. Similarly, to make the system work, humans had to exercise agency over matter and systematically manipulate the things of creation, for it was only through this type of experimentation, proceeding according to an orderly plan, that the rules and laws of nature could be discovered. The twin assumptions of human agency and freedom of choice infuse all of Bacon's discussions of method and procedure.

The ascription of agency and a co-operative role to humankind in the Instauration had the significant implication that something of the course and direction of sacred history was placed in the hands of people, even if God, in his omniscience, had already accounted for it. This is evident in the *Valerius Terminus* when Bacon, in acknowledging that vanity must be the ultimate end of all human works, also allows that the coming of the fourth age might be delayed through proper human effort: "vanity must be the end in all human effects, eternity being resumed, though the revolutions and periods may be delayed."[35] Bacon says no more of this possibility, but it is an idea that is entirely compatible with his belief in human free will. In the interactions between the free-willing God and His free-willing creatures it was not counter to the faith that human actions might actually have an effect on outcomes. As this providential age was one of mercy, it was reasonable that, if it were going well, it might be prolonged. However, it is also possible that the end of the world might be prolonged for negative reasons: if people chose not to act on the opportunity for Instauration which was before them, the golden age might be delayed until later generations understood what they were to do.

After his impeachment in 1621, Bacon's optimism about the imminence of the Instauration waned along with his political fortunes.[36] He began speaking of the Instauration as something far off, even if it had a fitful beginning in his own age. Although Bacon had always expected the Instauration to take more than one man's lifetime,[37] after his impeachment his hopes and his intellectual bequest were more frequently placed in the hands of future generations, rather than his own. In his 1622 *Historia Naturalis et Experimentalis*, dedicated not to the king but to his heir, Prince Charles, Bacon lamented that his own age preferred "to walk on in the old path, and not by the way of my Organum."[38] The proper path had been opened to the people of his own period, but they had chosen not to follow it. Bacon explained that, in doing so, they chose to repeat the act of hubris which caused the fall rather than recover from it:

---

[35]   WFB, vol. III, p. 222.

[36]   The theme of how Bacon's view of the Instauration changed after 1621 is explored more fully in Steven Matthews, "Francis Bacon's Scientific Apocalypse" in Cathy Gutierrez and Hillel Schwartz (eds), *The End that Does: Art, Science, and Millennial Accomplishment.* (London, 2006), pp. 93–111.

[37]   See Bacon's discussion of the Instauration timetable in *Valerius Terminus*: "That although the period of one age cannot advance men to the furthest point of interpretation of nature, (except the work should be undertaken with greater helps than can be expected), yet it cannot fail in much less space of time to make return of many singular commodities towards the state and occasions of man's life" (WFB, vol. III, p. 250).

[38]   WFB, vol. V, p. 133. Latin, WFB, vol. II, p. 15.

For we copy the sin of our first parents while we suffer for it. They wished to be like God, but their posterity wish to be even greater. For we create worlds, we direct and domineer over nature, we will have it that all things are as in our folly we think they should be, not as seems fittest to Divine wisdom, or as they are found to be in fact; and I know not whether we more distort the facts of nature or our own wits; but we clearly impress the stamp of our own image on the creatures and works of God, instead of carefully examining and recognizing in them the stamp of the Creator himself. Wherefore our dominion over creatures is a second time forfeited, not undeservedly; and whereas after the fall of man some power over the resistance of creatures was still left to him – the power of subduing and managing them by toil and arts – yet this too through our insolence, and because we desire to be like God and follow the dictates of our own reason, we in great part lose.[39]

Bacon goes on to encourage his readers to rethink, and make the right choice. After the fall, the freedom Adam had to choose between the design of God and his own pride was still possessed by Adam's heirs, although they had misused it so far. Bacon's Patristic turn on the issue of free will not only facilitated his explanation of sacred history and possibly his method itself, but also provided a mechanism for explaining why, if the Instauration had been decreed by God, it was failing to occur. Eventually God's prophecy would win out, but by human will the times and seasons could be delayed.

There are many possible errors which could result in the delay of the Instauration. Humanity might cling to the old and mistaken natural philosophy of Aristotle, and thus never enter upon the "new way." People might also simply not recognize the possibility of recovery, and therefore make no move to effect it. But there was, for Bacon, a more subtle and pernicious error that had been made in the past, and he has warnings about it throughout his writings on natural philosophy: the error of confusing the two books – the book of scripture and the book of nature – and reading one as if it were the other.

The Problem of Confusing the Two Books

Bacon addresses the error of confusing the two books in Aphorism 65 of the first book of the *Novum Organum.* This is the oft-quoted passage where he argues against an improper "admixture of theology" in natural philosophy, and concludes that "we be sober-minded and give to faith that only which is faith's."[40] Unfortunately, this passage is usually quoted to make a point that Bacon never made – that there should be a strict separation between science, or the study of the natural world, and theology. Given everything else Bacon says about the matter, this would be a very odd demand for him to make. For example, Bacon's definition of "natural theology" in the later work, *De Augmentis Scientiarum,* is "that knowledge concerning God, which may be obtained by the light of nature and the contemplation of his creatures."[41] In the words of John Henry, in such passages as Aphorism 65 "Bacon was not so much concerned that science and religion should not be mixed, but that they should not be

---

[39]　WFB, vol. II, p. 14; translation, WFB, vol. V, p. 132.

[40]　WFB, vol. I, pp. 175–6; translation, WFB, vol. IV, pp. 65–6.

[41]　WFB, vol. I, p. 544; translation, WFB, vol. IV, p. 341.

mixed the wrong way."[42] To understand Bacon's true intent in Aphorism 65 we must first pay attention to his original Latin, and, second, look at passages where he makes this same point more clearly.

Bacon is issuing his warning in connection with very specific examples from the Greeks and from certain moderns. Between the two sets of examples he denounces all such errors as "*errorum Apotheosis*," the "apotheosis of error," or the divinization of error. Bacon is objecting to the error of idolatry which, for him, is always a matter of replacing the truth with fantasies of human construction, whether in religion or natural philosophy. The condemnation here stands between two parallel structures, each beginning with *Huic...*, but referring to different kinds of error as if they were essentially the same. The first type of error is the error of the Greeks, and particularly Pythagoras, who used natural philosophy to construct religious and metaphysical ideas. The Greeks have attempted to derive the revealed truth of religion, the will of God, out of nature. The second type of error is that of certain theologians of Bacon's day who attempted to derive natural philosophy out of the source of spiritual truth – that is, the Scriptures, and Genesis or the book of Job, in particular. Either is idolatry, by Bacon's definition, and the errors, though on opposite sides, amount to a confusion of the two books. This fits precisely with what Bacon regarded as "heresy" in his *Meditationes Sacrae* – either ignorance of the will of God revealed in the Scriptures, or ignorance of God's power, which is revealed in the creatures. Looking in one book for what is revealed in the other will always lead to error.

Specifically in regard to those who seek the principles of natural philosophy in the Scriptures, Bacon parodies the language of Luke 24:5, where, in viewing the empty tomb after Jesus' resurrection, the women disciples are asked by an angel, "*Quid quaeritis viventem cum mortuis?*" or "Why do you seek the living among the dead?". Bacon accuses those who look for keys to the material order among the Scriptures of "*inter viva quaerentes mortua*," or seeking the dead among the living. In reversing the word order he makes it clear that he regards the Scriptures as the book which gives life, and is oriented toward the resurrection narrative of which this verse is a part, not toward the understanding of dead matter. Bacon's meaning here can be understood far better from his earlier, and more specific, discussion of the same topic in *The Advancement of Learning*:

> But for the latter [the philosophical understanding of the Scriptures], it hath been extremely set on foot of late time by the school of Paracelseus, and some others, that have pretended to find the truth of all natural philosophy in the Scriptures; scandalizing and traducing all other philosophy as heathenish and profane. But there is no such enmity between God's word and his works. Neither do they give honour to the Scriptures, as they suppose, but much imbase them. For to seek heaven and earth in the Word of God, whereof it is said, Heaven and earth shall pass, but my word shall not pass, is to seek temporary things amongst the eternal: and as to seek divinity in philosophy is to seek the living amongst the dead, so to seek philosophy in divinity is to seek the dead amongst the living: neither are

---

[42]   John Henry, *Knowledge is Power: How Magic, the Government and an Apocalyptic Vision Inspired Francis Bacon to Create Modern Science* (London, 2002), p. 86.

the pots or lavers whose place was in the outward part of the temple to be sought in the holiest place of all, where the ark of the testimony was seated.[43]

The scriptural reference to Luke 24:5 is set forth here in a manner which makes it clear that Bacon is primarily interested in preserving the proper priority of divinity over philosophy. The reference to the structure of the Old Testament temple reveals how Bacon regarded natural philosophy and divinity as both necessarily parts of a single priestly service. The distinction between the subject matter of the two books, as it is here set forth, is clearly not a distinction of the sacred and the secular, but a distinction between degrees of holiness. The subject of the book of nature is equated with the outer parts of the temple, and divinity is equated with the holiest place – the place of the mysterious dwelling of God in unapproachable cloud, seated upon the Ark. The study of nature is still sacred, but it deals with those things which are approachable; it cannot reach into the *Deus Absconditus*. By contrast, the study of divinity deals with things unapproachable by human reason, but revealed. The clergy of the Church attended to the holiest part of the temple, while Bacon's new order of natural philosophers were given care of the outer courts. The system was interactive, for, although natural philosophy could not pass into the mysteries of God, it could lead the mind up to the point where the mysteries of God could be properly contemplated. This passage is consistent with Bacon's belief that natural philosophy entailed a quasi-mystical ascent of the mind. In Book 1 of *The Advancement of Learning* he argued that natural philosophy leads back to religion, for the human mind passes from the things of nature to providence, and to meditation on that which is beyond second causes, which are the causes observable in nature.[44]

The same distinction between revealed and empirical knowledge is made in the *Valerius Terminus*, where the original of the phrase, "render unto faith only the things that are faith's" is found:

> Nay further, as it was aptly said by one of Plato's school the sense of man resembles the sun, which openeth and revealeth the terrestrial globe, but obscureth and concealeth the celestial; so doth the sense discover natural things, but darken and shut up divine. And this appeareth sufficiently in that there is no proceeding in invention of knowledge but by similitude: and God is only self-like, having nothing in common with any creature, otherwise as in shadow and trope. Therefore attend his will as himself openeth it, and give unto faith that which unto faith belongeth ...[45]

The will of God was secret and could not be known unless it was directly revealed. This belonged to faith. The study of the natural world could add nothing to human understanding of the hidden things of God, for it was impossible that nature could reveal anything of God's essence or will; it only revealed His power, and the pattern in which that power operated. As Bacon explained the same concept in *The Advancement of Learning*, the knowledge of nature could produce nothing but wonder in regard

---

[43] WFB, vol. III, p. 486.

[44] Ibid., p. 268.

[45] WFB, vol. III, p. 218.

to the knowledge of God.[46] It is important to note that in this statement Bacon meant by the knowledge of God the knowledge of God's secret will, and his transcendent identity, for, of course, knowledge of God's power did come through nature, and the witnessing of this power drew the mind of man upward through the chain of causes until the dependence of all things upon God was recognized.[47]

If the distinction between the subject matters of the two books is properly retained the result will be a well-grounded natural theology. This is another key feature of the Instauration event. Although God's will in the saving action in the incarnation had become well known, the understanding of His power had been incomplete because the book of nature had not been properly read before. In *De Augmentis Scientiarum*, Bacon presented his definition of natural theology in the context of the very distinction between the two books which we have just observed:

> For Natural Theology is also rightly called Divine Philosophy. It is defined as that knowledge, concerning God, which may be obtained by the light of nature and the contemplation of his creatures; and it may truly be termed divine in respect of the object and natural in respect of the light. The bounds of this knowledge, truly drawn, are that it suffices to refute and convince Atheism, and to give information as to the law of nature; but not to establish religion.[48]

Bacon continued, in this section, to reiterate the error of the heathen in this regard and then to list what can and cannot be known according to natural theology, thus elucidating the boundaries of his distinction:

> And therefore therein the Heathen opinion differs from the sacred truth; for they supposed the world to be the image of God, and man the image of the world; whereas the Scriptures never vouchsafe to attribute to the world such honour as anywhere to call it the image of God, but only the work of his hands; but man they directly term the image of God. Wherefore that God exists, that he governs the world, that he is supremely powerful, that he is wise and prescient, that he is good, that he is a rewarder, that he is an avenger, that he is an object of adoration – all this may be demonstrated from his works alone; and there are many other wonderful mysteries concerning his attributes, and much more touching his regulations and dispensations over the universe, which may likewise be reasonably elicited and manifested from the same; and this is an argument that has by some been excellently handled. But on the other side, out of the contemplation of nature and elements of human knowledge to induce any conclusion of reason or even any strong persuasion concerning the mysteries of faith, yea, or to inspect and sift them too curiously and search out the manner of the mystery, is in my opinion not safe. "Give unto faith the things which are faith's."[49]

Bacon had no small role for the study of nature in theology, and he has also clearly laid out, here, just what came under the heading of the "power of God." That this passage is analogous to Aphorism 65 is clear not only from the use of the phrase, "Give unto faith the things which are faith's," but also from his conclusion that,

---

[46]   Ibid., p. 267.
[47]   Ibid., p. 268.
[48]   WFB, vol. I, p. 544; translation, WFB, vol. IV, p. 341.
[49]   WFB, vol. I, p. 245; translation, WFB, vol. IV, pp. 341–2.

if the boundary is crossed, what will result is "at once an heretical religion and an imaginary and fabulous philosophy" [*religionem haereticam ..., et philosophiam phantasticam et superstitiosam*] – words that are also used in Aphorism 65.[50] Coming three years after the *Novum Organum*, this passage demonstrates that Bacon was consistent in his concern over the confusion of the two books from his earliest treatment in the *Meditationes Sacrae* to the last. Natural theology would finally be properly established through the Instauration, but this could only happen when the distinction between the two books was recognized and maintained.

According to Bacon, when humanity finally read both books correctly they would find themselves in the midst of the long prophesied age of recovery, when mastery over nature would produce an age of wonders and the relief of human suffering. The *New Atlantis* offers us a glimpse into what Bacon envisioned regarding this final age of the world. Guided by Christian charity, the right religion, and the proper reading of the book of nature, the Bensalemites live in a society of peace where the inhabitants lack no good thing. Disease and suffering have been controlled by wondrous foods, medicines, and special baths. Human flight has to some degree been accomplished. Special breeds of plants, animals, and fishes have been developed for the service of humanity. Most significantly, the wonders produced in Salomon's House are not yet at an end. Experimentation and prayer continues, with the result that new discoveries are constantly being made.

The Possibility of Immortality

The constant advance of natural philosophy on Bensalem raises the question of just how far the recovery of edenic power would go, in Bacon's thinking. As we observed in Chapter 3, there are key texts which state that he regarded Adam's knowledge as completely recoverable through hard work and mental labor. As a result of Genesis 3:19, hard work would always be necessary, but, beyond this qualification, it is evident that Bacon entertained the possibility of a complete recovery of paradise in the final age of the world. This is evident in how he deals with the possibility of immortality in the golden age of the Instauration. In the *Valerius Terminus* Bacon defined the task ahead of the natural philosophers as "a discovery of all operations and possibilities of operations from immortality (if it were possible) to the meanest mechanical practice."[51] When he described the parts of the curse which were "peremptory and not to be removed," he included the necessity of labor (now that creation had become rebellious), and the fact that all human efforts would be cut short by the second coming.[52] On the curse of death Bacon is notably silent.

It is clear from Bacon's *History of Life and Death*, written in 1623, that he regarded the immortality of Eden as a possibility. He wrote: "Whatever can be repaired gradually without destroying the original whole is, like the vestal fire, potentially eternal."[53] The error of previous generations of physicians was principally that they

---

[50] WFB, vol. I, pp. 245–6; translation, WFB, vol. IV, p. 342.

[51] WFB, vol. III, p. 222.

[52] Ibid., pp. 222–3.

[53] WFB, vol. V, p. 218; Latin, WFB, vol. II, p 106.

had taken too narrow a focus on this complex question, concerning themselves only with the loss of the body's original moisture. Bacon observes that the body repairs itself even in old age, but "[i]n declining age repair takes place very unequally."[54] One key to the solution is to balance out the physical process of repair, but, as with everything for Bacon, there is also a spiritual component to the question.

Bacon draws support for the program of extending human life from the evidence of the early Church. He wrote in justification of the pursuit of long life: "Besides, the beloved disciple survived the rest, and many of the Fathers, especially holy monks and hermits, were long lived."[55] Bacon's reverence for monks and hermits emerges again, for they stand alongside Saint John as examples of how piety and longevity can coincide. It is not surprising, in this light, that those with significantly long lifespans in Bensalem are also the hermits.[56] Long life is important because of Christian charity. As Bacon remarks, "they who aspire to eternity set little value on life," but "even we Christians should not despise the continuance of works of charity."[57] Bacon does not mention the monastics he has in mind by name, but that is really not necessary. Many monastics were known in the hagiographical writings for their longevity as well as their piety, but the paradigm for these hagiographies is always Athanasius' *Life of St Antony*.

Antony, as Athanasius describes him, lived to the age of one-hundred and five, and died only when his work on behalf of his followers was through. Remarkably, when Antony died he was in perfect health, but he knew that the time for him to move on to the next stage of his existence had come.[58] This, Athanasius tells us, is a death worthy of imitation, no less than Antony's life was worthy of imitation.[59] Bacon's ideal is very close to that set forth by Athanasius: a life spent in piety and in charity is ended because it is the proper time to pass from time to eternity. According to Augustine, this was the original design of life in the Garden of Eden as well.

Both the Eastern and Western Church Fathers describe the innocent state of humanity in the Garden as one which was not static but of implied growth. The Eastern Fathers, however, were not in the habit of asking the hypothetical question of how humans would have passed from mortality to immortality if Adam had not sinned. This was a question raised by the Pelagians and so answered, naturally, by Augustine. According to Augustine, the passage would still occur for Adam, but, "although he had a natural and mortal body, he should have in it a certain condition, in which he might grow full of years without decrepitude, and, whenever God pleased, pass from mortality to immortality without the medium of death."[60] That the East generally concurred with this judgment can be seen in the later summary of John of Damascus regarding Adam: "For being intermediate between God and matter he was destined, if he kept the command, to be delivered from his natural relation to existing

---

[54] Ibid.
[55] Ibid., p. 217.
[56] WFB, vol. III, p. 157.
[57] WFB, vol. V, p. 217.
[58] NPNF series 2, vol. 4, p. 221.
[59] Ibid., p. 219.
[60] NPNF series 1, vol. 5, p. 16.

things and to be made one with God's estate, and to be immovably established in goodness."[61] For none of the Fathers did this transformation entail leaving creation behind as a negative thing. Rather, as John of Damascus implies, upon becoming immortal, the human's relationship to creation would change, as humanity would gain the same perspective regarding creation as that held by the Creator.

In Bacon's age of Instauration, when Eden would be recovered, the passage between one life and the next would still be a reality, but it might be rendered a predictable and painless one. We may still call this transition "death," as it involves the separation of soul from body, as it did for Antony, but the uncertainty and the tragedy of this transformation might be removed. This certainly raises questions about the appropriateness of humans controlling the time and manner of the passage, but these are questions which Bacon does not answer. Presumably such decisions, like the Instauration itself, would be a matter of divine and human co-operation. What is clear is that Bacon was entertaining the idea of potentially restoring immortality to humanity through the course of the Instauration, thus making it a genuine recapitulation of Eden, as was the deathless, final golden age of Irenaeus, also.

Although it is tempting to cast Bacon's vision for the "Autumn of the World" as a radical innovation in theology, it is far more appropriate to describe it as a unique compromise. In a society where radical and literal millennialism was gaining ground alongside a host of more moderate apocalyptic notions, Bacon's was clearly one of the more moderate ideas. If his concept owes much to the theology of Irenaeus, it is also a tempering of it. What Irenaeus had not seen clearly, Bacon could, and the reality of the fulfillment of prophecy was less extreme than the ancient Church Father had assumed. The millennium of Irenaeus has been transferred to a golden autumn of what Bacon defined in the *Confession of Faith* as the third age of creation. There would be no fires and cataclysms, and no reign of the Antichrist other than what may have already occurred. There would be no dubious resurrection of the righteous and no earthly reign of Christ prior to the end. These features, all of which marked the more radical millennial systems, were absent from Bacon's golden age, which was, by comparison, far more orthodox. It bore the marks of catholic tradition in the context in which this tradition was being recast by men such as Lancelot Andrewes. It differed from Andrewes' eschatology only in its optimism regarding human knowledge. From Bacon's perspective there was nothing innovative about the Instauration. Prophecies which were once strange were visibly being fulfilled in his own day. This had been the plan from the beginning of the age.

---

[61]   NPNF series 2, vol. 9, p. 44b.

Chapter 6

Bacon's Circle and his Legacy

Bacon's vision of the Instauration reflected his own historical context. What subsequent generations would do with that vision would always reflect theirs. This is why we have such a proliferation of different interpretations of Bacon today. Within his own lifetime his literary circle – those who were most acquainted with his writings – tended to share his high regard for Patristics, the Catholic tradition, and his anti-Calvinism. Puritans were notably absent and Calvinists were scarce among his literary colleagues. A generation later those embracing his program most zealously were Puritans of a strongly millennial bent. Later still, the architects of the real-world version of Salomon's House would look to him, at least in their rhetoric, as one who effectively separated the scientific enterprise from divisive religious questions. The Enlightenment would shift Bacon's reputation still more. By the end of the Enlightenment he would be denounced by some as a source of modern atheism. This is truly an ironic turn, given Bacon's original motivation for the improvement of natural philosophy. Parts of this story have been told thoroughly by others. What remains is to provide a profile of Bacon's literary circle, and to set the parts of the story in order, so that we may understand how the interpretation of Bacon has changed when divorced from his original context.

Bacon's Literary Circle

Bacon's list of correspondents is long, but there were very few with whom he discussed his program for the reform of natural philosophy. During the early, formative years of his writings on natural philosophy only Lancelot Andrewes and his close friend Tobie Matthew figure prominently as confidants and editors of the Instauration texts. This privacy concerning his work reflects the political caution we have seen in Bacon's 1609 letter to Andrewes. Bacon always had an eye on his political fortunes, and it is evident from the manner in which he discusses his Instauration program, when it does come up, that he was careful not to allow his ambitions regarding philosophical reform to affect his political standing. This is not to say that Bacon's political ambitions were more important to him than his philosophical reform. If anything, the reverse was true. As he made clear in his 1592 letter to his uncle, Lord Burghley, his ambition for political position served his real interest of the reform of natural philosophy.[1] Bacon needed political office in order to have the means to publish his new method and put his reforms into practice. Whilst he took his legal work and his service to the Crown seriously, it was in the reform of

---

[1]   WFB, vol. VIII, pp. 108–9.

learning that he anticipated his lasting legacy. As he wrote of his reform of learning in the *Proemium* of the *Instauratio Magna*, "Certain it is that all other ambition seemed poor [literally, 'inferior'] in his eyes compared with the work which he had in hand."[2] If his program was to go forward, then he could not afford to lose the political position which made it possible. Yet he was proposing something in his Instauration writings that challenged established authorities and ran counter in its theology to the Calvinist majority. He needed to introduce it with caution, and only after those he trusted most had given their input.

After he began publishing, others were added to his list of assistants. He had a series of secretaries including a John Young, Thomas Meautys, William Rawley, Thomas Bushell, and Thomas Hobbes. Of these, the majority were fiercely loyal to their master and his memory, Hobbes being a notable exception. Thomas Bodley was consulted for his opinion at one time, but the exchange ended in conflict. The heated correspondence between the two is significant for our understanding of how Bacon's work was initially received by a *bona fide* Puritan of the early Stuart reign. Other occasional assistants in Bacon's literary endeavors included: John Selden, John Burough, Henry Wotton, George Herbert, Arthur George, William Boswell, and Bacon's brother in law, John Constable.[3]

Little is known about the religious convictions of some of Bacon's literary associates. Some, on the basis of a lack of any more specific evidence regarding their beliefs, would be best characterized simply as supporters of the king and the national Church.. Meautys, Borough, George, and Boswell are all presented in this light in the *Dictionary of National Biography*. Of John Young, Bacon's secretary, little is known beyond his service to Bacon, both as secretary and as an executor of his estate.[4] There is also little information available on John Constable. At any rate, with the exception of Bacon's loyal secretaries Meautys and Young, these names had little to do with Bacon's literary production. With regard to Borough, George, Constable, and Boswell there is no evidence of long-term work with Bacon on the Instauration. John Borough apparently helped Bacon obtain texts after his impeachment and exile from the verge made it difficult to obtain sources in London.[5] In 1619 Arthur George translated *de Sapientia Veterum* into English and the *Essays* into French, but he does not appear to have worked with Bacon beyond that. Constable and Boswell were chosen by Bacon as his literary executors and were charged with binding and publishing his works. The choice of these two makes sense because of the family connection with Constable and Boswell's extensive publishing connections. We know that both men were seriously remiss in carrying out their duties, leaving others,

---

[2]   WFB, vol. IV, p. 8. Cf. the Latin, WFB, vol. I, p. 122: "*Certe aliam quamcunque ambitionem inferiorem duxit re quam pre manibus habuit.*"

[3]   Extant correspondence was the key to establishing Bacon's literary circle. For a more thorough discussion of the circle, see Steven Paul Matthews, "Apocalypse and Experiment: The Theological Assumptions and Religious Motivations of Francis Bacon's Instauration" (unpublished dissertation, University of Florida, 2004), Chapter 3, pp. 160–229.

[4]   See WFB, vol. XIV, p. 229.

[5]   The best record of how Borough assisted Bacon is in Daniel Woolf, "John Selden, John Borough, and Francis Bacon's History of Henry VII, 1621," *Huntington Library Quarterly*, 47/1 (1984), pp. 47–54.

such as William Rawley, to see to the actual publication of Bacon's manuscripts. The remaining names in the circle include those with whom Bacon worked most closely on his Instauration writings, and we know enough about them to draw some important conclusions regarding religion and the first generation of "Baconians."

Tobie Matthew (1577–1655)

Apart from Andrewes, the most significant of Bacon's literary colleagues was Tobie Matthew. Sir Tobie Matthew was Bacon's trusted and close friend – arguably his closest friend – from the time when the two first met. This occurred around 1601 when Matthew came to London as a member of parliament for Newport in Cornwall.[6] Bacon wrote his essay "On Friendship" in response to a special request by Matthew.[7] This was telling, for the essay deals with the value of honest advice, and having someone with whom to share all things in good times and adversity.[8] Their friendship lasted through Matthew's recusancy and Bacon's impeachment, and Matthew was intimately involved with advising and assisting Bacon on the Instauration program throughout. When he was in England Matthew served as Bacon's personal courier and representative, and when he was on the continent he saw to the revision, translation, and publication of Bacon's works there. The correspondence between Matthew and Bacon is extensive from those periods when Matthew was away from England, but rather scant otherwise, which supports the idea that Bacon preferred to handle matters pertaining to the Instauration in person.

Tobie Matthew was born in Salisbury in 1577, the son of a clergyman by the same name. His father would later become bishop of Durham and then archbishop of York.[9] Like Bacon, Matthew did not remain in the faith to which he was born. By his own account, the younger Tobie never shared his father's staunch Reformed convictions, but it was still a shock to Bacon and many others when Matthew returned in 1607 from three years on the continent and revealed that he had converted to the Roman Catholic faith. For Matthew, this was the only reasonable decision, as Church history and tradition formed a thoroughly compelling argument against the innovations of the Reformation, and this was augmented by his fascination with the mystical spirituality of the Roman tradition.[10] Francis Bacon was the first person in whom he confided upon his return.[11] Matthew was imprisoned for several months in the Fleet for his recusancy, during which time many clergy remonstrated with him, including Lancelot Andrewes.[12] However, Matthew remained firm in his Roman Catholicism. Eventually, the efforts of Bacon and others secured his release, and in

---

[6] G. Walter Steeves, *Francis Bacon: A Sketch of his Life, Works and Literary Friends* (London, 1910), p. 200.

[7] WFB, vol. XIV, p. 429.

[8] WFB, vol. VI, pp. 437–43.

[9] DNB, vol. XIII, p. 63.

[10] See Matthew's letter to the nun, Dame Mary Gage, as found in Arnold Harris Matthew, *The Life of Sir Tobie Matthew, Bacon's Alter Ego* (London, 1907), pp. 79–80.

[11] WFB, vol. XI, p. 8.

[12] Ibid., p. 9; Matthew, *The Life of Sir Tobie Matthew*, pp. 89–92.

1608 he went into exile on the continent. In 1614 he was ordained to the priesthood, and he did not return to England until 1617. Matthew's writings are almost entirely theological, and geared toward the defense of the Roman Catholic Church as the true, historical Christian Church.

It has been recently established that Tobie Matthew was the author of a pro-Catholic tract entitled *Charity Mistaken.*[13] This work is an answer to the charge of certain Protestants in England that the Roman Catholics were uncharitable because they refused to admit that Protestants might be saved. Matthew responded that it was precisely on this point that the Roman Catholic Church was entirely charitable, because it was true that there was no salvation outside of the jurisdiction of the pope, and the Catholic Church earnestly desired that the Protestants should return. The particular opinion against which he was arguing had been presented by Bishop Lancelot Andrewes in a discussion between the two men many years earlier. As Matthew remembered the conversation, Andrewes had told him:

> … that he held the English Protestant Catholic Church, and the Roman Catholic Church, to be one and the same Church of Christ, forasmuch as he might conceive the fundamental points of faith, and the substantial worship and service of God; that we were both ... the same house of God; and that the only question between us both was, in very deed, and might justly be, whether that part of the house wherein they dwelt, or else that other part which we inhabit, were the better swept, and more cleanly kept, and more substantially repaired.[14]

Andrewes and Bacon could fault Matthew for moving backward into the "superstition" (as Bacon called it)[15] of the papacy, but he had not abandoned his Christianity. Matthew was genuinely concerned for the salvation of those, such as Bacon and Andrewes, who were outside the Roman communion but still respected Christian tradition and the opinions of the Fathers. From Matthew's perspective, if such Protestants were to understand the Scriptures and the Fathers properly, they would join him in his decision.

Matthew also included discussions of natural philosophy in his theological writings. His book, *Of the Love of our Only Lord and Savior, Jesus Christ*, contains a discussion, in very Baconian fashion, of how God has expressed his power in the laws of nature which can be observed in the visible realm. The study of nature is inherently a devotional exercise, according to Matthew's presentation, for God "by creation of the world ... led men up, by meanes of visible things, toward a knowledge, and beliefe of the invisible."[16] We may note the similarity of this statement to the

---

[13] Tobie Matthew, *Charity Mistaken, with the want whereof, Catholickes are uniustly charged for affirming, as they do with grief, that Protestancy unrepented destroies Salvation* (London, 1630). For the revised ascription to Sir Tobie Matthew, not, as hitherto, to Matthew Wilson, see A.F. Allison, "Sir Tobie Matthew, the Author of Charity Mistaken, *Recusant History*, 5 (1959), pp. 128–30.

[14] From Matthew's own account of his conversion, as quoted in Maurice F. Reidy, S.J., *Lancelot Andrewes: Jacobean Court Preacher* (Chicago, 1955), p. 82.

[15] See Bacon's letter to Matthew on the subject in WFB, vol. XI, p. 10.

[16] Tobie Matthew, *Of the Love of our Only Lord and Saviour, Jesus Christ* (Antwerp, 1622), p. 234.

common passages in Bacon's philosophical writings where knowledge is described as ascending until it breaks off in wonder at the contemplation of God.[17] In the preface, Matthew used the regular laws of nature as a pattern for spiritual learning and growth, with the understanding that God's works always follow the patterns which He has established. Throughout the first chapter, Matthew drew upon the analogies of observable nature for understanding the power of God. In another significant passage, he discussed the importance of the study of nature by the Magi, who were led to salvation as a result of their many years of contemplation of the stars:

> For as they had much imployed themselves, upon the contemplation of nature, by meanes of the Starrs; so by a starre, which was the likelyest lure to which they might be drawne to stoope, (for though their eyes looked upward for a while, yet soone after, it brought them downe upon their knees, at the sight of the divine infant) he vouchsafed to summon them to his serious.[18]

Written in 1622, at the same time that Matthew was working on the Latin of certain portions of Bacon's *De Augmentis Scientiarum*, the connection to Bacon's theology is unmistakable.

William Rawley (1588–1667)[19]

William Rawley was Bacon's personal chaplain, but he also assisted in his experiments and observations. After Bacon's death Rawley became his first biographer and his *de facto* literary executor. He was responsible for posthumously publishing the *Sylva Sylvarum* and the *New Atlantis*, translating Bacon's English works into Latin, and producing the earliest collections of Bacon's correspondence and literary remains. Rawley made Bacon's acquaintance sometime around 1612, and in 1616 Bacon was instrumental in obtaining for Rawley the rectorship of Landbeach. When Bacon became Lord Chancellor in 1618, he chose Rawley as his personal chaplain to replace William Lewis, for whom Bacon had recently obtained the position of provost at Oriel College. (William Lewis, for his part, was noted in the *Dictionary of National Biography* to be a "zealous member of the high-church party."[20]) From that time on, Rawley worked extensively with Bacon on all of his philosophical writings. Rawley received his doctorate in Divinity in 1621. Despite opportunities to advance in the Church, Rawley chose to remain in service to Bacon until the latter's death in 1626, after which he was appointed chaplain to both Charles I and Charles II in their respective reigns. At the very least, his views were compatible with the anti-Calvinism that had triumphed in the court after 1625.

---

[17] See *The Advancement of Learning*, Book 1, WFB, vol. III, p. 301. Such contemplation was the business of man's intellect in the original state: see p. 296. See also *Valerius Terminus*, WFB, vol. III, p. 218.

[18] Matthew, *Of the Love of our Only Lord and Saviour*, p. 84–5.

[19] For the basic outline of Rawley's life, see DNB, vol. 16, pp. 767–8. Many of the extant manuscripts of Bacon's works and observations from the years of Rawley's service to Bacon are in the hand of Rawley, indicating the extent to which Bacon used him as an amanuensis, as well as Rawley's suitability for seeing to the posthumous publication of the literary remains.

[20] DNB, vol. 11, p. 1078.

Rawley's only surviving theological work is one published sermon, preached at Easter 1623. It is based on Matthew 5:5, "Blessed are the meek for they shall inherit the earth." The printed sermon was dedicated to Bacon, and Rawley appears to have had his master in mind during its composition. A central theme of the sermon is that the meek suffer nobly when they are brought low by their enemies, which, given that the sermon was given after Bacon's impeachment, is likely meant for his benefit as much as any. While later historians might find it a stretch to judge Bacon as "meek", in his biography Rawley described Bacon's "long-suffering," reservation in speech and passions, and unwillingness to speak ill of anyone (key elements in the definition of meekness in this sermon) as his defining virtues.[21]

The sermon appealed to many different authorities in a manner which Puritans opposed, as Bacon had noted in his *Advertisement Touching the Controversies of the Church of England*. In Rawley's sermon Scripture never stands alone. In interpreting the text of the sermon Rawley made free use of a multitude of Church Fathers and Catholic opinions, from Saints Cyprian, Ambrose, Gregory, Dionysios, and Chrysostom, to Bernard and the Scholastic theologians. Seneca also makes an appearance as a model of classical virtue. In this blend of sources, as well as his use of careful semantic analysis, Rawley's homiletical approach is strongly reminiscent of Andrewes' sermons.

A key element of "meekness" for Rawley was the virtue of "golden mediocrity," which was embedded in his patron's family motto. From Rawley's discussion we may gain a certain insight into what the meaning of *mediocria firma* would have been to one who was very close to Bacon. "Mediocrity," notably, did not mean compromise, but referred to the classical ideal of a balanced life, with reason in control of the passions at all times.[22] The proper mediocrity required that anger must be both warranted, and "bridled by reason."[23] The same was true for anger's opposite (but less destructive) number – happiness.

Rawley concluded his sermon with the reward which would be granted to the "meek:" the inheritance of the earth. On this point Rawley presented Bacon's own theological motivation for the Instauration. The promise of "inheriting the earth" was properly to be understood as twofold. Rawley explained that the acts of charity carried out by the "meek" in this life would lead not only to "inheriting the earth" in the next life, when a "new heaven and a new earth" would be established, but also to a genuine inheritance of the earth in this life, just as Job saw (and Rawley's choice of word is significant) the "Instauration of his happiness."[24] By incorporating Bacon's concepts of charity and earthly recovery into his sermon, Rawley demonstrated that he himself did not regard the deeply religious language Bacon employed in discussing the purpose of his work as mere rhetoric: it was a sincere statement of the religious motivation for the Instauration.

---

[21]   Consider the definition of meekness in pp. 5–9 of William Rawley, *A Sermon of Meeknesse* (London, 1623). See also the description of Bacon's virtues in the *Life of Bacon*, WFB, vol. I, pp. 14–15.

[22]   Ibid., pp. 6–7.

[23]   Ibid., p. 34.

[24]   Ibid., pp. 53–5.

Henry Wotton (1568–1639)[25]

Henry Wotton was Bacon's kinsman, and the two had been together in the service of the Earl of Essex. Both escaped sharing Essex's downfall, and they continued as friends and correspondents until Bacon's death.[26] Wotton was a poet, and wrote the inscription on the headstone Thomas Meautys placed in St Michael's church in memory of Bacon. Bacon had apparently shared some of his own verse with Wotton from time to time,[27] but the two also shared an interest in Bacon's *forte* – natural philosophy.

Wotton dabbled in experiments in medical distillations and the measurement of time. As a diplomat he made numerous trips to the continent where he gathered all the information he could on current experiments. While Bacon was composing the *Novum Organum*, Wotton provided him with a written account of experiments he had witnessed some years earlier in the house of Johannes Kepler in Linz. When the book was completed, Bacon sent Wotton three copies of the *Novum Organum* on the understanding that Wotton would distribute them while he was in Germany to some of his contacts in natural philosophy.[28] Wotton promised to send one to Kepler.

From 1624 onward Wotton was provost of Eton College, where he was devoted to his pedagogical duties. He was noted for his piety and was ordained a deacon in 1627, a development he regarded as important for his own personal sense of vocation and for the benefit of his students at Eton.[29] He wanted to model the proper religion for them, which, as he associated it with the wearing of the surplice, was that of the "high-church" faction. In his own words, he hoped:

> … that gentlemen and knights' sons, who are trained up with us in a seminary of Churchmen (which was the will of the holy Founder) will by my example (without vanity be it spoken) not be ashamed, after the sight of courtly weeds, to put on a surplice.[30]

Wotton was a good friend and admirer of William Laud, and sided with the ascendant Arminians against the Calvinists. he spent some hours every day after chapel, "reading the Bible and authors in Divinity."[31] Among other devotional material, he was in possession of a manuscript copy of Bacon's *Confession of Faith*, which he held in high regard.[32] The influence of Bacon's *Confession* may be reflected in

---

[25] The most complete account of Wottton's life and activities is the two-volume "life and letters" biography of Logan Pearsall Smith, *The Life and Letters of Sir Henry Wotton* (2 vols, Oxford, 1907). The account presented here, unless noted otherwise, is summarized from these volumes as well as DNB, vol. 21, pp. 966–72.

[26] Bacon's nephew, Edmund Bacon, was Wotton's closest friend, and he married Wotton's niece, Philippa Wotton, adding a dimension to the familial ties between Francis Bacon and Henry Wotton. See Smith, *Life and Letters of Henry Wotton*, vol. 2, pp. 460–1.

[27] "Specimens of Bacon's poetry were also found among Wotton's papers after his death, and these were subsequently published in the *Reliquiae Wottonianae* in the year 1651" (Steeves, *Francis Bacon*, p. 218).

[28] WFB, vol. XIV, p. 131.

[29] Smith, *Life and Letters of Henry Wotton*, vol. 1, pp. 202–3.

[30] Ibid., vol. 2, p. 305.

[31] Ibid., vol. 1, p. 211.

[32] He passed it on, with praise, to Bacon's nephew, Edmund. See ibid., vol. 2, p. 393.

the reference to Christ as mediator in Wotton's last will and testament: "My Soul I bequeath to the Immortal God my Maker, Father of our Lord Jesus Christ, my blessed Redeemer and Mediator, through His all and sole sufficient satisfaction for the sins of the whole world, and efficient for His Elect."[33] The last phrase is an anti-Calvinist formula akin to one used by Arminius. It establishes that Christ did not die merely for some predestined "Elect," but that those who choose to be among the "Elect" will reap the benefit.[34]

Thomas Bushell (1594–1674)[35]

Thomas Bushell entered Bacon's household service in 1609 at the age of 15. In addition to his other duties as a servant, he assisted Bacon by taking notes on various experiments, which, as he recalled in his later writings, had much to do with metals. Bushell also engaged in his own experiments, which on more than one occasion incurred serious debts, which Bacon paid off for him. Bushell remained in Bacon's service continuously until Bacon's death, with the exception of a brief intermission at the time of Bacon's impeachment. Later, beginning in 1636, Bushell went on make a fairly successful career in mineral speculation. By introducing new methods of mining and mineral extraction, he succeeded in renovating and improving the royal mines at a number of locations. In the Civil War Bushell was an ardent Royalist, and held Lundy Island for King Charles I until the king allowed him to surrender it in February 1647. Bushell then went into exile until 1652, when he was allowed to return. In 1658 the Lord Protector granted him the right to work the old royal mines again, and, with the Restoration of the monarchy, Bushell continued his mining under Charles II. Thomas Bushell was a lifelong admirer of his old master, and in his writings he frequently referred to the influence of Bacon's instruction and example in his life.

Bushell's writings are filled with quotations and anecdotes from his years with Bacon, and his subsequent career as an innovator in mining certainly reflects the vision of his master. On matters of Bacon's life, however, Spedding concluded that he was "a bad authority at best."[36] Spedding's judgment is well founded, for Bushell had a remarkable tendency for remembering events and conversations that never could have happened. At one point in his writings Bushell recorded a lengthy speech that he insisted Bacon had prepared for the House of Lords, in which he would discuss with them his plan to erect, in Britain, the "Solomon's House" which he had "modelled" in his *New Atlantis*.[37] This speech would necessarily have been prepared around 1620, for, according to Bushell, Bacon's plans were interrupted by his impeachment. However, although the speech is remarkably detailed, it could never have been written

---

[33]   Ibid., vol. 1, p. 215.

[34]   See James Arminius, *Works of James Arminius,* trans. W.R. Bagnall, (Auburn and Buffalo, 1853), vol. 3, pp. 409–10, 438, and 458–9.

[35]   See DNB, vol. 3, pp. 487–9.

[36]   WFB, vol. XIV, p. 199.

[37]   Bushell recorded the speech in *An Extract by Mr. Bushell of his late Abridgement of the Lord Chancellor Bacons Philosophical Theory in Mineral Prosecutions* (London, 1660). Postscript pagination, pp. 18–19.

by Bacon. Among other problems, the *New Atlantis* did not see the light of day until Rawley published it in the year after Bacon's death, and the speech is written with the assumption that the Lords are all familiar with this work. Bushell's creative memories may be a function of his own self-fashioning, for he used this speech to establish the idea that King James had given Bacon control of certain flooded mines prior to the Civil War, thereby giving a pre-war royal pedigree for his own work in these same mines. On the other hand, Bushell was, beyond doubt, eccentric.

Nowhere was Bushell's eccentricity more apparent than in his piety, which also bore the mark of Bacon's religious values. Bushell's writings, as well as his life, evince a curious fascination with asceticism and the hermitical life, which he always ascribed to the wise influence of Bacon. As we have seen in Bacon's writings, there may be some grounding for Bushell's claim. Bushell wrote that Bacon advised him to take up the life of a hermit, spending his time in prayer and fasting, and thereby gain control of his sensual appetite before moving on to the higher work of natural philosophy. Then he could put Bacon's theories concerning minerals into practice:

> But he suddenly falling from an eminent height, as I by that time had deviated from his grave directions in the secure Paths of Vertue, imposed on me a new task, which was, not to search the Rocky bosoms of the barren Mountains, but, by a timely retirement to some solitary place where I might seclude myself from the treacherous vanities of the tumultuous world, to explore the deceitful Meanders of my stony heart, and when Divine grace should have assisted my better Reason in overcoming the rebellious affections of my Sensual appetite, if then the like Providence should call me thence to a more active life in the prosecution of his mineral documents, I should without any regret of my former penance attend the good hand of God in that design with humble patience; assuredly believing, that since he had supported me in the conquest of my self, he would conduct me through all difficulties, to the accomplishing so great a work for my Countryes good, and his own glory.[38]

Whatever Bacon actually said to Bushell on the subject, it is evident from his account here that the younger man took it very much to heart. Bushell used the ascetic life as a preparation for the work of natural philosophy. After Bacon's death, he spent three years on a desolate island in the Irish Sea, living alone in a hut 470 feet above the water, and on a scant diet of "herbs, oil, mustard, and honey, with water sufficient."[39] After this episode, he returned to the mainland, and sought his fortune in the mines. Bushell recounted Bacon's advice in, among other places, two addresses to prisoners during the time of the Commonwealth. He exhorted the prisoners to work in the mines as an ascetic discipline, claiming that hard work and the scant diet of the experience would help them develop the proper penitent attitude for salvation.[40]

John Selden (1584–1654)

It is difficult to say exactly when John Selden met Francis Bacon. Upon Bacon becoming Lord Chancellor, Selden composed *A Brief Discourse Touching the Office*

---

[38] Ibid., pp. 30–31.
[39] DNB, vol. 3, p. 488; and Bushell, *An Extract by Mr. Bushell*, p. 31 and postscript.
[40] Bushell, *An Extract by Mr. Bushell*, pp. 29–41.

of Lord Chancellor of England, suggesting some familiarity already in 1618.[41] After Bacon's impeachment in 1621 Selden helped him obtain texts for his various projects, since Bacon could not enter London himself.[42] We know that Selden performed this service regarding the *History of the Reign of King Henry the Seventh*, but whether Selden assisted with the philosophical works is less certain. A letter from Selden to Bacon, written in 1621, is particularly cordial, and suggests that the two men frequently discussed Bacon's projects, and that Selden's assistance was a matter of routine.[43] An early draft of Bacon's will lends support to the idea that the two had a close working relationship. He instructed John Constable to consult with Selden, as well as with an unidentified "Mr. Herbert of the Inner Temple" regarding which of his works should be published.[44]

We do know that Bacon's thought influenced Selden's learning and method. David Berkowitz has drawn attention to the many ways in which Selden was clearly Baconian in his approach to history, having drawn his principles of historical method from the principles for sound scientific method in the *Instauratio Magna*. He also had a notable interest in natural philosophy, and followed Bacon rejection of abstract principles in favor of observation.[45]

A lawyer by profession, Selden was highly regarded as a historian and a linguist, and not without cause: he had mastered Hebrew, Aramaic, Greek, Syriac, Arabic, and other languages, and employed them in his extensive historical studies. By 1640 Selden had an established reputation as "one of England's foremost orientalists."[46] Already in 1605, Selden had done a study of the Syrian mythology found in the Old Testament, *De Diis Syris*, which was well received throughout Europe after its publication in 1617.[47] Although his interests as a scholar ranged widely, it is clear from his writings that he had a special interest in Old Testament Hebrew and the Church Fathers. For him, Christian antiquity, but, even more, Jewish antiquity, held a place of special authority which always trumped the tenets of Reformation theology. Selden always measured recent developments in religion against ancient norms, which he identified through a rigorously critical reading of the ancient sources.

John Selden has a history of being viewed as a man lacking in piety and religious conviction. This is quite possibly attributable to the fact that he never hesitated to criticize or reject any opinions that he found to lack scholarly merit, even when they were held by eminent theologians and bishops. Selden's faith came into question numerous times in his own lifetime. As he was actively writing from the reign of King James through the early years of the Commonwealth, and frequently weighed

---

[41]   David Sandler Berkowitz, *John Selden's Formative Years: Politics and Society in Early Seventeenth-Century England* (Washington, 1988), p. 35.

[42]   Woolf, *John Selden* ..., pp. 47–54.

[43]   Ibid., p. 52.

[44]   WFB, vol. XIV, p. 540.

[45]   Berkowitz, *John Selden's Formative Years*, pp. 32–3, 48, 69. A more direct discussion of the interrelationship of Bacon's induction and Selden's development of a concept of natural law is presented by Reid Barbour, *John Selden: Measures of the Holy Commonwealth in Seventeenth-Century England* (Toronto, 2003), pp. 179–87.

[46]   Berkowitz, *John Selden's Formative Years*, p. 41.

[47]   Ibid., p. 40.

in on matters pertaining to the faith, it is not surprising that charges flew. In his case they came from all sides.

One of those who denounced Selden as "being more learned than pious" was Sir Simonds D'Ewes, who was himself a man of "pronounced puritanical views."[48] There was good reason for a Puritan to be bothered by Selden's opinions. Selden opposed the Westminster Assembly's addition of Calvinist confessional documents, such as the Lambeth Articles or the results of the Synod of Dort, as a doctrinal standard for the Church of England.[49] Selden placed a strong emphasis on human free will, and this led him to ridicule the Calvinist understanding of predestination.[50] On the other hand, Selden also angered many of the bishops of England when he argued in his scholarly tract, *On the History of Tithes*, that the English tithe system could not be defended historically or scripturally.[51] Although Selden was brought up on charges for this tract, King James did nothing by way of punishment.[52] An influential bishop who had the king's ear broke ranks with his fellow bishops, and defended Selden's scholarship as sound (even if his points did not apply to Stuart England). This bishop was Selden's friend, Lancelot Andrewes.[53]

Selden's own life and writings present an image of a man who was a critical thinker eager to test religious claims against historical "truth." As a result, he came up with both very traditional and very idiosyncratic answers to common questions. According to Selden: religion was to be found in the proper ceremony of the historic liturgy, not in the preaching of Puritans; the Scriptures were not to be read apart from the tradition of the Church; all religious texts fell into a hierarchy with the Jewish texts at the top, and Christian authorities, including the Church Fathers, next; and bishops were not absolutely necessary, but they were the most expedient way to run a church, and it was pointless to do away with a system that worked.[54]

At the end of his life Selden would embrace the tenets of Thomas Erastus, which subjected religion to the control of the state. However, the association of Erastianism with atheism or secularism (a common charge in the seventeenth century) is inaccurate in Selden's case. Selden's concept of placing state authority over matters of religion relied on the subjection of both Church and state to the one God, and is tied with his desire to reinstitute the Jewish Sanhedrin system as a means of government in England. Taken on his writings alone, John Selden was a man of profound convictions, with a high regard for tradition, church ceremony, and

---

[48] For D'Ewes' quotation itself see George W. Johnson (ed.), *Memoirs of John Selden and Notices of the Political Contest During His Time* (London, 1835), p. 362. For his Puritanism see DNB, vol. 5, pp. 901–2.

[49] Barbour, *John Selden*, pp. 164–5.

[50] See John Selden, *Table Talk* (London, 1689), p. 47, subject "Predestination."

[51] For a useful summary of the *History of Tithes*, and its implications, see Paul Christianson, *Discourse on History, Law, and Governance in the Public Career of John Selden, 1610–1635* (Toronto, 1996), pp. 68–83.

[52] Berkowitz, *John Selden's Formative Years*, p. 36.

[53] Florence Higham, *Lancelot Andrewes* (London, 1952), p. 88.

[54] These points are all made in Selden's *Table Talk*, published posthumously.

the Church Fathers, although he measured all things according to his own canon of practicality and reverence.[55]

George Herbert (1593–1633)

The poet George Herbert was corresponding with Bacon by 1620, and was evidently courting Bacon's patronage at that time.[56] It is possible that the two men had met earlier, through their mutual friend, Lancelot Andrewes, or through another mutual friend, Henry Wotton.[57] In the early 1620s Herbert assisted Bacon by translating sections of *The Advancement of Learning* into Latin to be used in the more extensive work, *De Augmentis Scientiarum.*[58] It is clear from several poems that Herbert wrote in Bacon's honor, including a memorial after his death, that Herbert held Bacon in high regard.[59] Bacon, for his part, dedicated his *Translation of Certaine Psalmes into English Verse* to Herbert with the following words:

> The pains that it pleased you to take about some of my writings I cannot forget; which did put me in mind to dedicate to you this poor exercise of my sickness. Besides, it being my manner for dedications, to choose those that I hold most fit for the argument, I thought that in respect of divinity and poesy met, (whereof the one is the matter, the other the stile of this little writing,) I could not make better choice. So, with signification of my love and acknowledgment, I ever rest
>
> Your affectionate Friend,
Fr. St. Alban[60]

In this way Bacon acknowledged his gratitude for Herbert's work on the *De Augmentis*, and dedicated a work to him, which reflected an interest they shared in the poetic reinterpretation of the book of Psalms.[61] This was by no means the only interest which the two men shared. Herbert took Bacon's plan for the Instauration seriously, and contributed to the propagation of Bacon's ideas.

In the early 1630s Herbert collaborated on an English translation of three tracts published under the name of the first and longest work, Leonard Lessius' *Hygiasticon*.

---

[55]   Selden's plan for the restoration of the Sanhedrin is a key theme in Reid Barbour's book. For the other features of Selden's theology see the discussion in Matthews, "Apocalypse and Experiment", pp. 199–218.

[56]   Joseph H. Summers, *George Herbert: His Religion and Art* (Binghamton, NY, 1954), p. 40.

[57]   For Herbert's association with Andrewes see ibid., p. 30; and Paul Welsby, *Lancelot Andrewes* (London, 1964, p. 108. On Herbert's long friendship with Wotton see Izaak Walton, *Life of George Herbert* (London, 1670), p. 27.

[58]   Summers, *George Herbert*, p. 32. Summers also discussed the significant influence which Bacon, and Baconian themes, had upon Herbert's poetry and language.

[59]   Ibid., p. 40.

[60]   WFB, vol. VII, p. 275.

[61]   On Herbert and the Psalms see Chana Bloch, *Spelling the Word: George Herbert and the Bible* (Berkeley, 1985), pp. 233–6. Bacon's activity in the *Translation of Certaine Psalmes*, according to Bloch, was far from an isolated phenomenon. According to Izaak Walton, Andrewes also dedicated a collection of his genuine translations of the Psalms out of Hebrew to Herbert. See Walton, *Life of George Herbert*, p. 26.

All of the works deal with dietary directions for the "preserving of life and health to extream old age."[62] (Ironically, this book was published in 1634, the year after Herbert's death at the age of forty.) Those involved in the book's production saw themselves as engaged in a single collaborative project, and evidently were in common discussion on matters of diet, and how moderation in diet could lengthen and improve life. The justification for the book was the example of "The Late Viscount St. Albans," Francis Bacon, who had given his own directions for the prolongation of life, and had entrusted those who came after him with the task of building upon his work. The book begins with an extract from Bacon's *History of Life and Death*. The works which follow it are clearly understood as an elaboration upon Bacon.

George Herbert's theology is a matter of ongoing, and at times heated, debate among scholars.[63] There are certain points which must be agreed upon by all, and in considering two points of this scholarly consensus we can get a sense of how Herbert fits into the pattern of the Bacon circle. First, on the question of predestination, and its correlative doctrines of election and reprobation, Herbert was unquestionably Calvinist. Predestination, for Herbert, was double, it was absolute, and it had nothing to do with human choice,[64] and he had engaged Andrewes in a lively, but friendly, debate on this very issue.[65] Nonetheless, Herbert also believed in a genuinely free human will in things that were "outward," or not pertaining to spiritual matters.[66] Second, in his liturgical affinities, Herbert tended toward positions which were epitomized in his day by Andrewes and Laud, and which would later come to be associated with the "high-church" movement.[67] As Gene Edward Veith has pointed out, however, this reflected a liturgical conservatism common among Protestants, and should be seen as separating Herbert from Puritans and Presbyterians, and not necessarily from Calvinism, particularly as Calvinism was received in England by Cranmer and others.[68] Nevertheless, in an England already polarized over ceremony, Herbert's position necessarily put him into a "camp."

Among the members of Bacon's literary circle who remained on good terms with Bacon, Herbert is the only recognizable Calvinist. However, Herbert's Calvinism must always be qualified. In light of Bacon's theological shifts, it is notable that Herbert created a very un-Calvinist space for human free will. Herbert's defense of the received liturgical traditions of the English Church, as well as his general agreement with episcopal government, keep him out of any tidy association with Puritans or Presbyterians, though when Laudian reforms began to come into fashion

---

[62] Leonard Lessius, *Hygiasticon, or, The right course of preserving Life and Health unto extream old Age: Together with soundnesse and integritie of the Senses, Judgement, and Memorie* (Cambridge, 1634).

[63] See the summary of the field in Christopher Hodgkins, *Authority, Church, and Society in George Herbert* (Columbia, 1993), pp. 1–8. See also Gene Edward Veith, *Reformation Spirituality: The Religion of George Herbert* (Lewisburg, 1985), pp. 23–41.

[64] Hodgkins, *Authority, Church and Society*, pp. 16–20; Veith, *Reformation Spirituality*, pp. 83–116; Summers, *George Herbert*, pp. 57–8.

[65] Walton, *Life of George Herbert*, p. 26.

[66] Summers, *George Herbert*, p. 58.

[67] Ibid., pp. 55–7.

[68] Veith, *Reformation Spirituality*, p. 206.

toward the end of his life he did not associate with that movement either. Christopher Hodgkins has made a compelling case that Herbert could properly be labeled an "old Conformist," distinguishing him from the new conformity advocated by Laud, as well as from the more extreme reform movements. Herbert sought the fading conformity of the Elizabethan Settlement, as represented by Whitgift and others.[69]

Thomas Hobbes (1588–1679)

Much can be learned from those members of Bacon's literary circle who admired Bacon and incorporated his vision for the Instauration into their own writings. Much can also be learned by considering those members of the circle who did not. Thomas Hobbes was one who did not look back on Bacon with affection. An appeal to a connection with Hobbes would play well into the portrayal of Bacon as an atheist, or at least as a skeptic. There can be no doubt that Hobbes assisted Bacon for some period of time, but, beyond this, any connection between the two men becomes problematic. A.P. Martinich, the recent biographer of Hobbes, issued a strong caution on making too much of the relationship between Hobbes and Bacon, because the differences between the two in their thinking, particularly in terms of natural philosophy, far outweigh the similarities.[70] Much the same could be said on matters of religion.

Thomas Hobbes may have known Bacon as early as 1614, and was working for him as a secretary and assistant by 1620.[71] Some connection appears to have persisted until Bacon's death in 1626. The earliest record we have of the nature of the relationship between Bacon and Hobbes is from John Aubrey's *Brief Lives*. The entry on Bacon came in 1681, late in the seventeenth century, and that which was not common knowledge was based entirely on Hobbes' personal recollections. Unsurprisingly, Hobbes appears "beloved" by Bacon in this account, but those parts which Hobbes has contributed to Aubrey's image of Bacon are not flattering, except to Hobbes. Bacon is portrayed as a distracted old man who would be lost without his most able secretary, Hobbes.[72] As Martinich has argued, there is little evidence that the two actually had anything more than a strict working relationship.

It may be that Hobbes was not so far from Bacon's position while he was working for Bacon as a youth. Hobbes' position on religion and Christian theology only became articulated when he was a member of the Great Tew Circle after 1636, ten years after Bacon's death.[73] According to Martinich, the "broad Socinianism," which even in the mature Hobbes "was married to an even more dominant fideism," was an evolutionary development, and in his earlier years, particularly around 1614, Hobbes was "a more conventional Protestant" than he would be later.[74] If Hobbes and Bacon were close at some point, it was before Hobbes had fully formulated his own views on religion and Christian theology, and it was only near the end of Bacon's life. If,

[69]   Hodgkins, *Authority, Church and Spirituality*, pp. 37–40.

[70]   A.P. Martinich, *Hobbes: A Biography* (Cambridge, 1999), p. 66.

[71]   Ibid., pp. 65–9, and 29.

[72]   John Aubrey, *Brief Lives* (London, 1950), pp. 9–16.

[73]   Martinich, *Hobbes*, pp. 104–6.

[74]   Ibid., pp. 34, and 106.

at one time, Hobbes learned from Bacon, he took what he learned in a radically different direction. As Martinich observed, "Bacon's radical empiricism in science is at the opposite pole from Hobbes' rationalism."[75] Bacon's maxim, "Knowledge is power," was consciously modified by Hobbes, who concluded "Knowledge is for the sake of power."[76] Bacon's knowledge was power and it served the holy end of charity. Hobbes's knowledge served the end of power.

Thomas Bodley (1545–1613)

Thomas Bodley is the only Puritan who, because at one point he did give Bacon feedback on a text, could be considered a member of Bacon's literary circle. Only one exchange between the two men on Bacon's philosophy is preserved in the correspondence, but it is particularly telling. In 1605 Bacon had contributed a copy of his *Advancement of Learning* to Bodley's library, among other choice recipients. It is evident from the letter accompanying the book that Bacon and Bodley were not particularly intimate, that the donation of the text to Bodley's library was *pro forma*, and that prior to this gift Bodley was evidently unaware of the nature of Bacon's philosophical interests.[77] Some time before 1608, Bacon included Bodley among his reviewers of a draft of a new work, entitled *Cogitata et Visa*, which contained a plain explanation of what Bacon regarded as wrong with the current system of learning and natural philosophy, as well as his plan for supplanting it with his own method. Bodley was slow to respond, and Bacon sent him a chastising letter, asking for his papers back, and declaring that Bodley was "slothful," and of no help.[78] Bacon suspected that Bodley disliked his argument, and it turned out that he was right. When Bodley finally did respond, his letter was long and relentlessly critical of Bacon's entire project.

Since, in Bodley's words, Bacon had included him among his "chiefest friends" by sending him the draft and asking for his comments on it, Bodley took the liberty of being brutally honest.[79] First, Bodley objected to Bacon's dismissal of the received authorities in natural philosophy and the present state of learning. Then he let Bacon know what he thought would be the inevitable result of abandoning their contemporary methods and starting afresh:

> … now in case we should concur, to doe as you advise, which is to renounce our common Notions, and cancell all our Actions, Rules, and Tenents, and so to come, Babes, *ad regnum naturae*, as we are willed by Scriptures to come, *ad regnum coelorum*, there is nothing more certain in my understanding then that it would instantly bring us to Barbarism, and after many thousand years, leave us more unprovided of theoreticall furniture, then we are at present … [80]

---

[75] Ibid., p. 66.

[76] Ibid., p. 276.

[77] WFB, vol. X, pp. 252–6.

[78] Ibid., 366.

[79] Bodley's entire letter is to be found in the 1648 collection of Bacon's "Remaines." See Tobie Matthew (ed.), *The Remaines of the Right Honorable Francis Lord Verulam …* (London, 1648), pp. 80–85.

[80] Ibid., p. 82.

Bodley continued in this vein at length, informing Bacon that he was unconvinced that Aristotle and the sciences at the dawn of the seventeenth century were in need of any reform. Ultimately, Bodley contended, there might be some new inventions and methods from time to time, but the knowledge of one learned age was little different from that of another. Just as Calvin concluded, the sciences could not be "perfected" as Bacon thought, or even much improved over their state in classical antiquity, "[f]or still the same defects that Antiquity found will reside in Mankind."[81] It was best not to abandon the insights of Aristotle once they had been recovered. Then, in the candor which Bodley presumed that Bacon would afford his "speciall friend," Bodley confided that Bacon would not find any positive reception for his material in the universities: "I stand well assured for the tenour and Subject, of your maine discourse, you are not able to impannell a substantiall Jury in any university that will give upp a verdict to acquite you of your errour."[82] We have no evidence of any further correspondence between Bacon and Bodley. It is reasonable to conclude that Bodley had presumed too much on the "friendship" of a man who had lost patience with him long ago.

As James Spedding concluded regarding the relationship between Bacon and Bodley:

> Bodley might help Bacon with supply of books; but for ideas, it must have been manifest from the moment his answer came that no light could be looked for from that quarter; not even the light which is given by intelligent opposition. Nothing can be weaker or more confused than his reasons for dissent, unless it be his apprehension of the questions at issue.[83]

Very soon after this exchange Bacon sent the same manuscript of the *Cogitata et Visa* to Lancelot Andrewes, a genuine friend who had always been far more open to Bacon's ideas. The pessimism Bodley expressed regarding the potential for human improvement on the sciences was a reflection of the Calvinist understanding of human nature after the fall. It was also a rejection of the theological justification for the Instauration.

Conclusions Regarding Bacon's Literary Circle

Those who worked most closely with Bacon during his lifetime were a diverse group, ranging from the Catholic priest, Matthew, to the moderate Calvinist, Herbert, to the original thinker, Selden. Many other members were clearly comfortable with the Laudian reforms when they occurred. The profile of the group as a whole, measured by their theological interests and priorities, fits well with the image of an emerging "high-church" Anglicanism, or Anglo-Catholicism, although the terms are anachronistic for Bacon's era. Anyone was welcome to join in the group effort of the recovery of learning, so long as they were qualified and accepted the theological premises behind the recovery.

---

81   Ibid., p. 83.
82   Ibid., p. 84.
83   WFB, vol. X, p. 366.

What the circle clearly did not include were Puritans. Given the dominance of Calvinism even among non-Puritans during Bacon's lifetime, there was a shortage of even moderate Calvinism in the Bacon circle. There is good reason for this. The founding principles of Bacon's Instauration run entirely counter to what Calvin asserted in the *Institutes*. Nowhere is this more apparent than on Calvin's doctrine of total depravity. Bacon's Instauration simply could not occur if the human intellect had become depraved and incapable of effecting its own recovery, or if, as Calvin put it, "soundness of mind and integrity of heart were withdrawn" as part of the punishment for sin.[84] With this in mind, it is particularly interesting to note that many in the next generation of Baconians were Calvinists.

The Reform of Learning in the Civil War and the Commonwealth

The story of the next generation of Baconians has been written most thoroughly by Charles Webster. In his book, *The Great Instauration: Science, Medicine and Reform 1626–1660*, Webster established that, in the years of the English Civil War and the Commonwealth, Bacon's vision of the Instauration became very popular among certain groups of England's intellectuals who joined together in associations dedicated to bringing Bacon's vision into being.[85] Webster has been criticized for using the term "Puritan" as a blanket label for those involved, when many of the names he mentions in the book would have been far from comfortable with the term. Although the criticism is valid, given the wealth of careful work that has been done to define "Puritan" according to historical context, the criticism cannot be allowed to eclipse the key points regarding religion and science raised by Webster's study.

One of Webster's central points is that religion and science were not separate pursuits in this generation of Baconians. As Webster observed, "science was dominated by men in holy orders (academics, chaplains, or beneficed); they covered the entire political and religious spectrum, from Beale, Wilkens, Wallis, Cudworth and More, to the radicals Dell, Webster, and Pinnel."[86] The literature produced by these circles, as exemplified by the correspondence of Samuel Hartlib, covers the gamut of human knowledge from religion, to politics, to natural philosophy. Webster also demonstrates that the millennialism of many of the leading intellectuals resonated with Bacon's own golden age. Bacon's interpretation of Daniel 12:4 as a prophecy of a new age of knowledge remained a central motivation for the explosion of experimentation and natural investigation which occurred during this time.[87] If Bacon's ideas were not taken straight, but blended with other sympathetic strains of thought, it still remains true that the second generation of Baconians recognized no separation of "science" and "religion" in Bacon's writing, and they generally made none in their own. The one significant exception to this rule is the group which began meeting in 1645 in London where, in the memory of John Wallis, "matters of

---

[84]   ICR, Book 2, ch. 2, sec. 12.

[85]   Charles Webster, *The Great Instauration: Science, Medicine and Reform 1626–1660* (New York, 1975).

[86]   Ibid., p. 39.

[87]   Ibid., pp. 1–31.

Theology and State Affairs" were not discussed, and the conversation was limited to natural philosophy due to the heated political climate of the Civil War.[88]

In light of our consideration of Bacon, another important point springs from Webster's book. Regardless of how strictly "puritan" any of the lead characters were in the second generation, they were overwhelmingly Calvinist. The political changes of the English Civil War were accompanied by a strong reaction against the anti-Calvinist Laudian reforms. From Laud's preference of Arminian clergy and professors, the pendulum moved back toward Calvinism under the Parliament. Calvinists held the positions that had the power to put Bacon's reforms in place, and Calvinists were the catalysts of the correspondence networks which ensured that knowledge "passed to and fro" for the increase of learning.

How Bacon's theology became acceptable to Calvinists is a question which may take another book or two to answer properly. One key to the problem is to recognize that groups such as "Calvinists" were simply not always and everywhere the same. The continental Calvinists who came to England and figure most prominently in Webster's narrative – Samuel Hartlib, John Comenius, and Henry Oldenburg – clearly did not share the strict interpretation of the Calvinist doctrines of total depravity and predestination espoused by the Puritans of Bacon's generation. Apparently, by this time, few among the English did either. As had been the experience of George Herbert, the Calvinist sympathies for Bacon's program, motivated as it was by piety and charity, had come to outweigh the differences which were so clear to Bodley and Bacon himself. It is also likely that those theological differences were not, by then, even clearly recognized, given the changing nature of theological discourse in the decades after Bacon's death. If Bacon was not a Calvinist, neither was he an Arminian, as defined by the conflict under Laud. The Calvinists driving the program of natural philosophy forward during the rule of parliament and the Protectorate could use Bacon's ideas without hesitation.

The Restoration and the Royal Society

The next chapter of Bacon's legacy concerns the generation which consciously saw itself as building Bacon's house of learning, Salomon's House, on English soil. Regardless of how Baconian they may or may not have been in actual method, the founders of the Royal Society presented themselves publicly as Bacon's heirs. The Society was, according to John Evelyn, "a design no way beneath that of his Solomons House."[89] Throughout Thomas Sprat's *History of the Royal Society of London* Bacon figures as the founder of those ideas which were put in practice by the Society. In the introductory poem to Sprat's *History* Abraham Cowley cast Bacon in the role of a latter-day Moses, a personal legacy which Bacon would have approved of:

---

[88]　　Christopher J. Scriba, "The Autobiography of John Wallis," *Notes and Records of the Royal Society of London*, 25/1 (1970), p. 40.

[89]　　Quoted in William T. Lynch, *Solomon's Child: Method in the Early Royal Society of London* (Stanford, CA, 2001), p. 46.

From these and all long Errors of the way,
In which our wandering Praedecessors went,
An like th' old Hebrews many years did stray,
In Desarts but of small extent,
Bacon, like Moses, led us forth at last,
The barren Wilderness he past,
Did on the very border stand
Of the blest promis'd Land,
And from the Mountains Top of his Exalted Wit,
Saw it himself, and shew'd us it.[90]

In leading the people out of previous errors in natural philosophy toward a promised land of right philosophical method Bacon had been a prophetic figure, but, like Moses, Bacon had not lived long enough to experience what that promise would ultimately entail. There were similarities and differences between his vision and the reality of the Royal Society, and this was certainly true in the matter of religion.

Sprat's *History* contains clear parallels with Bacon's treatment of religion and natural philosophy. Sprat uses Bacon's own maxim to defend the idea that natural philosophy is not at odds with the faith, but rather a support for it: "That by a little knowledge of Nature men become Atheists; but a great deal returns them back again to a sound and Religious mind."[91] This conclusion follows from a very Baconian idea: that there cannot be conflict between religion and natural philosophy because natural philosophy is itself a form of religion. According to Sprat, natural philosophy was the religion of Adam in the Garden:

> This was the first service, that Adam perform'd to his Creator, when he obey'd him in mustring, and naming, and looking into the Nature of all the Creatures. This had bin the only Religion, if men had continued innocent in Paradise, and had not wanted a Redemption.[92]

Along with Bacon, Sprat insisted that, as a result of error, a twofold reformation had become necessary for revealed religion and natural philosophy, and this was taking place in England:

> From this I will farther urge, That the Church of England will not only be safe amidst the consequences of a Rational Age, but amidst all the improvements of Knowledge, and the subversion of old Opinions about Nature, and introduction of new ways of Reasoning thereon. This will be evident, when we behold the agreement that is between the present Design of the Royal Society, and that of our Church in its beginning. They both may lay equal claim to the word Reformation; the one having compass'd it in Religion, the other purposing it in Philosophy. They both have taken a like cours to bring this about; each of them passing by the corrupt Copies, and referring themselves to the perfect Originals for their instruction; the one to the Scripture, the other to the large Volume of the Creatures.[93]

---

[90] Thomas Sprat, *The History of the Royal Society of London* (London, 1667), introductory poem.
[91] Ibid., p. 351.
[92] Ibid., pp. 349–50.
[93] Ibid., pp. 362–3.

Much more may be said on the parallels between Sprat and Bacon, and there is evidence that Bacon's theological system had particular influence on the Royal Society's apologetics. Bacon's legacy was not merely preserved in the writings of Sprat, it was transformed. The differences are telling, especially as they reflect the changed historical situation of the Restoration.

Between Bacon and the Royal Society stood an age of civil war, which shaped the English thinking on matters of religion in ways that Bacon had not anticipated. Religion had not been progressively perfected, but rather had become implicated in a terrible conflict. The conclusion of Bacon's associate, John Selden, captured the disenchantment which reigned after the Civil War: "'Tis to no purpose to reconcile Religions, when the interests of princes will not suffer it. 'Tis well if they could be reconciled so far, that they should not cut one another's Throats."[94] The dream of a truly restored religion had been dashed in the minds of the majority, both Laudian and Puritan alike. The rigorous exclusion of theology, which, according to John Wallis,[95] was practiced during the heat of the Civil War, is not in evidence in Sprat's writings, but a post–Restoration attitude of tolerance is. Although Sprat retained Bacon's rhetorical flourish in insisting that the English Church stood at the apex of the reformation of religion, this apex was measured not by purity but by compromise.

The official policy of the Royal Society was to admit "[m]en of different Religions, Countries, and Professions of Life," which Sprat explains as a desire among the members "not to lay the Foundation of an English, Scotch, Irish, Popish, or Protestant Philosophy; but a Philosophy of Mankind."[96] There is certainly an echo of the openness and diversity of Bacon's own literary circle here, but there is also a difference. Bacon spent much time and energy in supplying his Instauration with a coherent theological foundation. Prophecy, exegesis, and salvation history contributed a theological framework to the Instauration which provided both the motivation and the justification for beginning the project of the reform of learning. Those who embraced his vision of Instauration would presumably do so because they embraced the ideas of the theological system behind it. Some of these ideas are clearly present in Sprat's *History*, as we have noted, but not the complete framework. The discussion of prophecy is notably absent, as is the discussion of an earthly golden age, and the careful consideration of the Genesis fall narrative.

Bacon believed in a refining and renovation of theology, as he expressed it in *De Augmentis Scientiarum*, which would lead toward a religion purged of error. His own contributions to exegesis and the "history of prophecy" were a part of that ongoing reformation. This belief is also absent in Sprat. The concern for a perfected theology has been replaced in the *History of the Royal Society* by an acceptable, basic doctrinal minimum which transcends religious differences and consists of what all Christians hold in common. The first part of this common core of Christian teaching is the "Evangelical Doctrine of Salvation by Jesus Christ,"[97] and the second part is recognizing God as the creator of nature, and accepting the resultant compatibility of the study of nature

---

[94]   Selden, *Table Talk*, p. 52.
[95]   Scriba, "Autobiography of John Wallis."
[96]   Sprat, *History of the Royal Society*, p. 63.
[97]   Ibid., p. 351.

and the Christian faith.[98] Then, according to Sprat, there are the basic doctrines which "have been long since deduc'd by consequences from the Scripture, and are now setled in the Body of that Divinity, which was deliver'd down to us by the Primitive Church, and which the generality of Christendom embraces."[99] Sprat is conveniently silent on what these universal doctrines are, having merely asserted that there are teachings which have always been embraced by the "generality of Christendom."

In the end, Sprat leaves religious conclusions in the hands of individuals. Regarding the Church of England and other denominations, Sprat concludes that "[i]t concerns them, to look to the reasonableness of their Faith; and it is sufficient for us, to be establish'd in the Truth of our own."[100] Although his language is guarded, a little further on Sprat entertains the idea that "all wise Men should have two Religions; the one, a publick, for the conformity with the people; the other, a private, to be kept within their own breasts."[101] Sprat makes no effort to draw a line between "science" and "faith," and, indeed, he concludes, like Bacon, that natural philosophy is a religious pursuit. However, Sprat also argues that Bacon's era was prevented from producing the house of learning of which he had dreamed, because it was too concerned with settling matters of religion:

> The Reign of King Iames was happy in all the benefits of Peace, and plentifully furnish'd with men of profound Learning. But in imitation of the King, they chiefly regarded the matters of Religion, and Disputation: so that even my Lord Bacon, with all his authority in the State, could never raise any Colledge of Salomon, but in a Romance.[102]

The religious divisions which had been so troublesome in recent history were to be tolerated and left to individual judgment, rather than overcome through continuing reform. Religious conclusions are a private matter, while natural philosophy is a public one. In this way, the groundwork for an effective separation of scientific and theological discourse has been laid. The Enlightenment would go on to develop this idea.

The Enlightenment Transformation of Bacon's Memory

After the hagiographical mention of Bacon by the early Royal Society the trail of Bacon's reception and legacy is harder to follow, particularly as his memory becomes implicated in the Enlightenment. In his recent popular survey of Bacon's thought and influence, *Knowledge is Power*, John Henry traced the idea that Bacon was an atheist, or at least a deist, to the Enlightenment readers, or misreaders, of Bacon, who saw his work as an anticipation of their own thinking: "Enlightenment thinkers wanted to see the heroic figures in the history of the new science as thinkers swayed only by rational and empirically grounded principles."[103] Henry's point has

---

[98] Ibid., p. 352.

[99] Ibid., p. 353.

[100] Ibid., p. 63.

[101] Ibid.

[102] Ibid., pp. 151–2.

[103] John Henry, *Knowledge is Power: How Magic, the Government and an Apocalyptic Vision Inspired Francis Bacon to Create Modern Science* (London, 2002), p. 83.

a great deal of merit, for it is certainly true that Bacon was never portrayed as being anything but a Christian prior to the Enlightenment. It is also undeniable that Bacon was canonized by many key figures in the Enlightenment, who went out of their way to portray him as one of their own, and they were often deists or atheists.

In Voltaire's *Lettres philosophiques*, Bacon is the "father of experimental philosophy," and hence the source of the advances over the past made in Voltaire's own age.[104] Among the Encyclopedists, d'Alembert presented Bacon as being responsible for ending the age of darkness which preceded him. Diderot regarded himself as a true disciple of Bacon, fashioned his own thought and writing after Bacon's, and gave Bacon credit for the idea and plan of the *Encyclopédie*.[105] It is not clear whether Bacon's Enlightenment readers were willing to portray Bacon *himself* as an atheist or a deist, even if his writings, as they interpreted them, led inexorably to their own positions.

Voltaire, for example, portrayed Bacon as the one who raised the scaffolding of "modern scientific thought," which got the whole enterprise of the study of nature moving in the right direction, but maintained that the scaffolding had been torn down now that Bacon's own philosophy has been superseded.[106] It appears that Voltaire regarded Bacon as very much susceptible to the whims of his own superstitious age, for he ends his discussion of Bacon by noting that Bacon's *History of the Reign of Henry the Seventh* was weakened by "this rigamarole that was formerly taken as inspired."[107] In adopting Bacon's basic plan for division of the sciences in the prospectus for the *Encyclopédie* Diderot preserved all of Bacon's categories, including "*Sciences de DIEU*," although this category received little space.[108] If Diderot was unwilling to do much with this category, he would nevertheless have been aware of Bacon's reasons for including it.[109]

Regardless of whether the *Philosophes* or Encyclopedists themselves regarded Bacon as an atheist, they were interpreted as doing so by one of the most vigorous early critics of the Enlightenment, Joseph de Maistre. De Maistre was eager to lay the blame for the errors of his own age on Bacon's method, and the motif of his *Examination of the Philosophy of Bacon*[110] is a condemnation of Bacon's ideas

---

[104] Voltaire, *Letters on England [Lettres philosophiques]*, trans. Leonard Tancock (New York, 1980), pp. 57–61.

[105] Cf. R. Loyalty Cru, *Diderot as a Disciple of English Thought* (New York, 1966). See p. 244 for crediting Bacon with the idea and plan of the *Encyclopédie*.

[106] Ibid., p. 58.

[107] Ibid., p. 62.

[108] See R.J. White, *The Anti-Philosophers: A Study of the Philosophes in Eighteenth Century France* (New York, 1970), p. 95.

[109] Bacon sets out his rationale for this category in Book Three of *De Augmentis Scientiarum*, and the terms of the discussion are undeniably theistic, and in accord with the general interests of the Church of England of his day. See WFB, vol. 1, pp. 539–44.

[110] Joseph de Maistre, *An Examination of the Philosophy of Bacon, Wherein Different Questions of Rational Philosophy are Treated* trans. Richard A. LeBrun (Montreal, 1998), pp. ix–xv. This was published in 1836, a full fifteen years after de Maistre's death. The criticism, therefore, can be associated with the Enlightenment, while the impact of this book, which has a much harsher treatment of Bacon than de Maistre's earlier *St. Petersburg Dialogues*, must be traced to the early to mid-nineteenth century.

which, he believed, had recently been taken to their logical conclusions and had proven disastrous to both throne and altar. Among de Maistre's ultimate criticisms was the charge that Bacon's method led to atheism, for there was no mistaking in Bacon's writings, "this concentrated hate, this incurable rancour against religion and its ministers, which has particularly distinguished most of the scholars and cultivated minds of our century."[111] Positive statements about religion in Bacon's writings were treated as hypocritical by de Maistre, who felt that he had more than enough evidence from Bacon's own writings, as well as from their impact on his own age, to recognize Bacon's real intent.

Whether the conclusion that Bacon was an atheist, or at least a critic of Christianity, had its origin in the Baconians of the Enlightenment or in their critic, Joseph de Maistre, a radically transformed image of Bacon emerged from the Enlightenment. Bacon's vision of an apocalyptic age, heralded by prophecy and ensured by providence, was lost. His theological statements were discounted as irrelevant or insincere. The historical Bacon had once more been refashioned in the image of his readers. The Christian philosopher had become the father of atheism.

Conclusion

The many different images of Bacon in contemporary scholarship have their precedents in the transformation of his legacy at the hands of subsequent generations of Baconians. Scholars still cast Bacon in the image of the Enlightenment deist or atheist, in the image of the Fellow of the Royal Society who separated the divisive issues of faith from natural philosophy, and in the image of a sincere Puritan. Behind all the Bacons of other ages and other agendas lies the historical Bacon of his own age and place. Here, in his own context, is found the young man rebelling against his mother's beliefs. Here is found the scholar of Scripture and Patristics, challenging the hegemony of Calvin and wrestling with the age-old question of providence versus free will. Here is found the visionary who, looking backward over sacred history, could suddenly perceive the divine pattern of it all, and looking forward, could see the age of plenty that God had promised. Here is found the friend of Andrewes and Matthew, seated amidst a constellation of some of the more original thinkers of the next generation. Here is found the man who saw his rise to Lord Chancellor as a means to a greater work and a higher calling. Here, in his proper context, is found the Francis Bacon whom we must recognize if we are to properly understand Bacon, his writings, and his legacy.

---

[111]   de Maistre, *Examination of the Philosophy of Bacon*, p. 271.

Bibliography

Primary Sources

Andrewes, Lancelot, Αποσπασματια *Sacra* (London, 1657).
———, *Works of Lancelot Andrewes*, ed. John Parkinson (11 vols, Oxford, 1854).
Aquinas, Thomas, *Summa Theologica*, ed. T.C. O'Brien (Cambridge: Cambridge University Press, 2006).
Arminius, James, *Works of James Arminius*, trans. W.R. Bagnall (Auburn and Buffalo: Derby Miller & Orton, 1853).
Augustine of Hippo, *De Civitate Dei*, in *Corpus Christianorum*, vols 47–8 (Turnholtii: Typographi Brepols, 1955).
Bushell, Thomas, *An Extract by Mr. Bushell of his late Abridgement of the Lord Chancellor Bacons Philosophical Theory in Mineral Prosecutions* (London, 1660).
Calvin, John, *A Commentarie of John Caluine, upon the first booke of Moses called Genesis*, trans. Thomas Tymme (London, 1578).
———, *Institutes of the Christian Religion*, trans. Ford Lewis Battles (Philadelphia: Westminster Press, 1960).
———, *Institutes of the Christian Religion*, trans. Henry Beveridge (Edinburgh, 1845).
———, *Institutio Christianae Religionis* (London, 1576).
Cartwright, Thomas, *A Treatise of Christian Religion, or, The Whole Bodie and Substance of Divinitie* (London, 1616).
Chemnitz, Martin, *Examination of the Council of Trent*, trans. Fred Kramer (4 vols, St Louis: Concordia, 1971).
Firth, Charles H. (ed.), *Stuart Tracts: 1603–1693* (Westminster: Archibald Constable and Co. Ltd, 1903).
Herbert, George, *The Works of George Herbert* (New York: Thomas Y. Crowell, 1880).
James, Thomas, *Catalogus Librorum Bibliothecae Publicae quam vir Ornatissimus Thomas Bodleius ...* (London, 1605).
Johnson, George W. (ed.), *Memoirs of John Selden and Notices of the Political Contest During His Time* (London: Orr and Smith, 1835).
Lessius, Leonard, *Hygiasticon, or, The right course of preserving Life and Health unto extream old Age: Together with soundnesse and integritie of the Senses, Judgement, and Memorie* (Cambridge, 1634).
Luther, Martin, *Luther's Works*, ed. Jaroslav Pelikan (54 vols, St Louis: Concordia, 1955–57).
Luther's predecessours, or, An answere to the qvestion of the Papists : where was your church before Luther? (London, 1624).
Maistre, Joseph de, *An Examination of the Philosophy of Bacon, Wherein Different Questions of Rational Philosophy are Treated*, trans. Richard A. LeBrun (Montreal and Kingston: McGill-Queen's University Press, 1998).

Matthew, Tobie, *Charity Mistaken, with the want whereof, Catholickes are uniustly charged for affirming, as they do with grief, that Protestancy unrepente destroies Salvation* (London, 1630).

———, *Of the Love of our Only Lord and Savior, Jesus Christ.* (Antwerp, 1622).

——— (ed.), *The Remaines of the Right Honorable Francis Lord Verulam ...* (London, 1648).

Melanchthon, Philip, *Loci Communes*, trans. J.A.O. Preus (St Louis: Concordia, 1992).

Migne, J.P. (ed.), *Patrologiae Cursus Completus, Series Graeca* (161 vols, Paris, 1857–66).

Perkins, William, *A Discourse of the Damned art of Witchcraft* in *The Workes of that Famous and Worthy Minister of Christ in the Universitie of Cambridge, Mr. William Perkins* (3 vols, London: I. Legatt, 1626–31).

———, *A Treatise of Man's Imaginations* (Cambridge, 1607).

———, *The Christian Doctrine* edition in English and Irish (Dublin, 1652).

Prado, Hieronymo, and Villalpando, Juan Bautista, *In Ezechiel explanationes et apparatus vrbis, ac templi Hierosolymitani* (Rome, 1596–1605).

Rawley, William, *A Sermon of Meeknesse* (London, 1623).

Roberts, Alexander, and Donaldson, James (eds), *The Ante-Nicene Fathers* (10 vols, Grand Rapids: Hendrickson, 1994). Reprint of the Edinburgh edition.

Schaff, Philip, and Wace, Henry (eds), *The Nicene and Post-Nicene Fathers* (2 series, 14 vols each, Grand Rapids: Hendrickson, 1994). Reprint of the Edinburgh edition.

Selden, John, *Table Talk* (London, 1689).

Spedding, James, Ellis, Robert Leslie, and Heath, Douglas Denon (eds), *The Works of Francis Bacon* (14 vols, London, 1858–74).

Sprat, Thomas, *The History of the Royal Society of London* (London, 1667).

Voltaire, *Letters on England [Lettres philosophiques]*, trans. Leonard Tancock (New York: Penguin, 1980).

Secondary Sources

Allison, A.F., "Sir Tobie Matthew, the Author of Charity Mistaken, *Recusant History*, 5 (1959), pp. 128–30.

Aubrey, John, *Brief Lives* (London: Secker and Warburg, 1950).

Barbour, Reid, *John Selden: Measures of the Holy Commonwealth in Seventeenth-Century England*, (Toronto: University of Toronto Press, 2003).

Barth, Karl, *Church Dogmatics*, trans. G.T. Thomson (Edinburgh: T&T Clark, 1936).

Bauckham, Richard, *Tudor Apocalypse: Sixteenth Century apocalypticism, millenarianism, and the English Reformation* (Oxford: Sutton Courtnay Press, 1977).

Bejczy, Istvan, *Erasmus and the Middle Ages: The Historical Consciousness of a Christian Humanist* (Leiden: E.J. Brill, 2001).

Berkowitz, David Sandler, *John Selden's Formative Years: Politics and Society in Early Seventeenth-Century England* (Washington, DC: Folger Books, 1988).

Bloch, Chana, *Spelling the Word: George Herbert and the Bible* (Berkeley: University of California Press, 1985).

Christianson, Paul, *Discourse on History, Law, and Governance in the Public Career of John Selden, 1610–1635* (Toronto: University of Toronto Press, 1996).

———, *Reformers and Babylon: English Apocalyptic Visions from the Reformation to the Eve of the Civil War* (Toronto: University of Toronto Press, 1978).

Collinson, Patrick, *The Birthpangs of Protestant England* (Houndmills, Basingstoke: Macmillan, 1988).

———, *The Elizabethan Puritan Movement* (Berkeley and Los Angeles: University of California Press, 1967).

Cross, Claire, *Church and People, 1450–1660: The Triumph of the Laity in the English Church* (Atlantic Highlands, NJ: Humanities Press, 1976).

Cru, R. Loyalty, *Diderot as a Disciple of English Thought* (New York: AMS Press, 1966).

Davis, J.F, "Lollardy and the Reformation in England" in Peter Marshall (ed.), *The Impact of the English Reformation: 1500–1640* (London: Arnold Press, 1997).

Dickens, A.G., *The English Reformation*, (2nd edn, University Park: Pennsylvania State University Press, 1991).

Firth, Katherine R., *The Apocalyptic Tradition in Reformation Britain: 1530–1645* (Oxford: Oxford University Press, 1979).

Florovsky, Georges, *The Collected Works of Georges Florovsky* (14 vols, Belmont, MA: Nordland Publishing Company, 1987).

Gallagher, P., and Cruikshank, D.W. (eds), *God's Obvious Design* (London: Tamesis, 1988).

Garrett, Christina H., *The Marian Exiles* (Cambridge: Cambridge University Press, 1938).

George, Charles, and George, Katherine, *The Protestant Mind of the English Reformation: 1570–1640* (Princeton, NJ: Princeton University Press, 1961).

Gonzalez, Justo L., *A History of Christian Thought* (3 vols, Nashville, TN: Abingdon Press, 1975).

Grant, Robert M., *Irenaeus of Lyons* (London: Routledge, 1997).

Grimm, Harold J., *The Reformation Era 1500–1650* (New York: Macmillan, 1954).

Hall, Basil, 'The Early Rise and Gradual Decline of Lutheranism in England 1520–1660" in *Studies in Church History,* Subsidia ii (Woodbridge: Boydell and Brewer, 1979).

Henry, John, *Knowledge is Power: How Magic, the Government and an Apocalyptic Vision Inspired Francis Bacon to Create Modern Science* (London: Ikon Books, 2002).

Higham, Florence, *Lancelot Andrewes* (London: SCM Press, 1952).

Hirsch, Emanuel, *Die Theologie des Andreas Osiander* (Gottingen: Vanderhoeck and Ruprecht, 1919).

Hodgkins, Christopher, *Authority, Church, and Society in George Herbert* (Columbia: University of Missouri Press, 1993).

Hotson, Howard, "The Historiographical Origins of Calvinist Millenarianism" in Bruce Gordon (ed.), *Protestant History and Identity in Sixteenth-Century Europe* (2 vols, Aldershot: Scolar Press, 1996).

Innes, David C., "Bacon's New Atlantis: the Christian Hope and the Modern Hope," *Interpretation*, 22/1 (1994), pp. 3–37.

Jacobs, Henry Eyster, *The Lutheran Movement in England* (Philadelphia: G.W. Frederick, 1890).

Jardine, Lisa, and Stewart, Alan, *Hostage to Fortune: The Troubled Life of Francis Bacon* (New York: Hill and Wang, 1998).

Jones, G. Lloyd, *The Discovery of Hebrew in Tudor England: A Third Language* (Manchester: Manchester University Press, 1983).

Kottman, Karl A. (ed.), *Catholic Millenarianism: From Savonarolla to the Abbé Grégoire* (Dordrecht: Kluwer Academic Publishers, 2001).

Lossky, Nicholas, *Lancelot Andrewes, the Preacher (1555–1626): The Origins of the Mystical Theology of the Church of England*, trans. Andrew Louth (Oxford: Clarendon Press, 1991).

Lossky, Vladimir, *The Mystical Theology of the Eastern Church* (Crestwood, NY: St Vladimir's Seminary Press, 1976).

Lynch, William T., *Solomon's Child: Method in the Early Royal Society of London* (Stanford, California: Stanford University Press, 2001).

McKnight, Stephen A., *The Religious Foundations of Francis Bacon's Thought* (Columbia: University of Missouri Press, 2006).

Martin, Julian, *Francis Bacon, the State, and the Reform of Natural Philosophy* (Cambridge: Cambridge University Press, 1992).

Martinich, A.P., *Hobbes: A Biography* (Cambridge: Cambridge University Press, 1999).

Matthew, Arnold Harris, *The Life of Sir Tobie Matthew, Bacon's Alter Ego* (London: Elkin Mathews, 1907).

Matthews, Steven Paul, "Apocalypse and Experiment: The Theological Assumptions and Religious Motivations of Francis Bacon's Instauration" (unpublished dissertation, Gainesville: University of Florida, 2004).

———, "Francis Bacon's Scientific Apocalypse" in Cathy Gutierrez and Hillel Schwartz (eds), *The End that Does: Art, Science, and Millennial Accomplishment* (London: Equinox Publishers, 2006), pp. 93–111.

———, "Reading the Two Books with Francis Bacon: Interpreting God's Will and Power" in Peter J. Forshaw and Kevin Killeen (eds), *The Word and the World: Biblical Exegesis and Early Modern Science* (Basingstoke: Palgrave Macmillan, 2007), pp. 61–77.

Meyendorff, John, *Byzantine Theology* (New York: Fordham, 1974).

Milner, Benjamin, "Francis Bacon: The Theological Foundations of 'Valerius Terminus,'" *Journal of the History of Ideas*, 58/2 (1997), pp. 245–64.

New, John F.H., *Anglican and Puritan: The Basis of Their Opposition, 1558–1640* (Stanford, CA: Stanford University Press, 1964).

Pelikan, Jaroslav, "Some Uses of Apocalypse in the Magisterial Reformers" in C.A. Patrides, and Joseph Wittrich (eds), *The Apocalypse in English Renaissance Thought and Literature* (Ithaca, NY: Cornell University Press, 1984), pp. 74–88.

———, *The Christian Tradition* (5 vols, Chicago: University of Chicago Press, 1978).

Peltonnen, Markku (ed.), *The Cambridge Companion to Bacon* (Cambridge: Cambridge University Press, 1996).

Quantin, Jean-Louis, "The Fathers in Seventeenth Century Anglican Theology" in Irena Backus (ed.), *The Reception of the Church Fathers in the West: From the Carolingians to the Maurists* (Leiden: E.J. Brill, 1997), pp. 951–86.

Quasten, Johannes, *Patrology* (4 vols, Westminster, MD: Christian Classics, 1984).

Reidy, Maurice F., S.J., *Lancelot Andrewes: Jacobean Court Preacher* (Chicago: Loyola University Press, 1955).

Ryrie, Alec, "The Strange Death of Lutheran England," *The Journal of Ecclesiastical History*, 53/1 (January 2002), pp. 64–92.

Schaff, Philip, *History of the Christian Church* (8 vols, New York: Charles Scribner's Sons, 1910).

Scriba, Christopher J., "The Autobiography of John Wallis," *Notes and Records of the Royal Society of London*, 25/1 (1970), pp. 17–46.

Scribner, Charles, *For the Sake of Simple Folk: Popular Propaganda for the German Reformation* (Cambridge: Cambridge University Press, 1981).

Smith, David L., "Catholic, Anglican, or Puritan? Edward Sackville, Fourth Earl of Dorset, and the Ambiguities of Religion in Early Stuart England" in Donna B. Hamilton, and Richard Strier (eds), *Religion, Literature, and Politics in Post-Reformation England: 1540–1688* (Cambridge: Cambridge University Press, 1996), pp. 115–37.

Smith, George, Stephen, Sir Leslie, and Lee, Sidney (eds), *Dictionary of National Biography* (21 vols, London: Oxford University Press, 1882–1900).

Smith, Logan Pearsall, *The Life and Letters of Sir Henry Wotton* (2 vols, Oxford, 1907).

Steeves, G. Walter, *Francis Bacon: A Sketch of his Life, Works and Literary Friends* (London: Methuen & Co. Ltd, 1910).

Stinger, Charles L., *Humanism and the Church Fathers: Ambrogio Traversari (1386–1439) and Christian Antiquity in the Italian Renaissance* (Albany, NY: State University of New York Press, 1977).

———, "Italian Renaissance Learning and the Church Fathers" in Irena Backus (ed.), *The Reception of the Church Fathers in the West: From the Carolingians to the Maurists* (Leiden: E.J. Brill, 1997), pp. 473–510.

Summers, Joseph H., *George Herbert: His Religion and Art* (Binghamton, New York: Center for Medieval Texts and Studies, 1954).

Trinkaus, Charles, *In Our Image and Likeness: Humanity and Divinity in Italian Humanist Thought* (2 vols, Chicago: University of Chicago Press, 1970).

Tyacke, Nicholas, *Anti-Calvinists: The Rise of English Arminianism, 1590–1640* (Oxford: Clarendon Press, 1987).

Veith, Gene Edward, *Reformation Spirituality: The Religion of George Herbert* (Lewisburg: Bucknell University Press, 1985).

Wallace, Karl, *Francis Bacon on the Nature of Man* (Champaign–Urbana: University of Illinois Press, 1967).

Walker, D.P., *Spiritual and Demonic Magic from Ficino to Campanella* (London: Warburg Institute, 1958).

Walton, Izaak, *Life of George Herbert* (London, 1670).

Webster, Charles, *The Great Instauration: Science, Medicine, and Reform 1626–1660* (New York: Holmes & Meier, 1975).

Welsby, Paul, *Lancelot Andrewes* (London: SPCK, 1964).

White, R.J., *The Anti-Philosophers: A Study of the Philosophes in Eighteenth Century France* (New York: Macmillan, 1970).

Whitney, Charles, *Francis Bacon and Modernity* (New Haven, CT: Yale University Press, 1986).

———, "Francis Bacon's Instauratio: Dominion of, and over, Humanity," *Journal of the History of Ideas*, 50/3 (1989), pp. 371–90.

Woolf, Daniel, "John Selden, John Borough, and Francis Bacon's History of Henry VII, 1621," *Huntington Library Quarterly*, 47/1 (1984), pp. 47–54.

Zagorin, Perez, *Francis Bacon* (Princeton, NJ: Princeton University Press, 1998).

Index